Making Taste Public

Also available from Bloomsbury:

Food Activism, edited by Carole Counihan and Valeria Siniscalchi
The Taste Culture Reader, edited by Carolyn Korsmeyer

Contents

BLOOMSBURY ACADEMIC
Bloomsbury Publishing Plc
50 Bedford Square, London, WC1B 3DP, UK

BLOOMSBURY, BLOOMSBURY ACADEMIC and the Diana logo are trademarks of Bloomsbury Publishing Plc

First published in Great Britain 2018

Cover design by Adriana Brioso
Cover image: © Tara Moore/Getty Images

A catalogue record for this book is available from the British Library.

Library of Congress Cataloging-in-Publication Data
Names: Counihan, Carole, 1948– editor. | Højlund, Susanne, editor.
Title: Making taste public : ethnographies of food and the senses / edited by Carole Counihan and Susanne Højlund.
Description: London, UK : Bloomsbury Academic, 2018. | Includes bibliographical references and index.
Identifiers: LCCN 2018009115| ISBN 9781350052680 (hb) | ISBN 9781350052673 (pbk.) | ISBN 9781350052697 (ePDF) | ISBN 9781350052703 (ePub)
Subjects: LCSH: Food habits—Social aspects—Case studies. | Food—Sensory evaluation—Social aspects—Case studies. | Taste—Social aspects—Case studies
Classification: LCC GT2855 .M28 2018 | DDC 394.1/2–dc23 LC record available at https://lccn.loc.gov/2018009115

ISBN: HB: 978-1-3500-5268-0
ePDF: 978-1-3500-5269-7
ePub: 978-1-3500-5270-3

Typeset by RefineCatch Limited, Bungay, Suffolk
Printed and bound in Great Britain

To find out more about our authors and books visit www.bloomsbury.com and sign up for our newsletters.

Making Taste Public

Ethnographies of Food and the Senses

Edited by Carole Counihan and Susanne Højlund

BLOOMSBURY ACADEMIC
LONDON • NEW YORK • OXFORD • NEW DELHI • SYDNEY

List of Figures

List of Contributors

Rachel E. Black is visiting Assistant Professor of Anthropology at Connecticut College. Her research focuses on food and wine, exploring issues of taste, sociability and work in Italy, France and North America. She is the author of *Porta Palazzo: The Anthropology of an Italian Market* (2010) and co-editor of *Wine and Culture: Vineyard to Glass* (2012).

Maria Carabello is a PhD student studying health policy and sociology at the University of Michigan. She holds a BS in Nutrition and Food Sciences and MS in Food Systems, both from the University of Vermont. Her research is broadly concerned with the interplay between food and health behaviors, population health trends, and public policy formation. Her scholarly research and reviews have appeared in *Appetite*, *Food, Culture & Society*, and the *Graduate Journal of Food Studies*.

Bodil Just Christensen is an Assistant Professor at the University of Copenhagen. She has done research on food agency and everyday eating as well as performative forms of eating, with a special focus on how eaters engage in food practices and other schemes of body management. She is currently part of an interdisciplinary research project that addresses the multiple factors potentially determining the large variation of weight loss after obesity surgery.

Carole Counihan is Professor Emerita of Anthropology at Millersville University and has studied food, gender, and culture in Italy and the USA for forty years. She is author of *A Tortilla Is Like Life* (2009), *Around the Tuscan Table* (2004), and *The Anthropology of Food and Body* (1999). She is co-editor of *Food and Culture: A Reader* (1997, 2008, 2013) and *Food Activism* (2014) and editor-in-chief of the scholarly journal *Food and Foodways*.

Peter I. Crawford is Professor of Visual Anthropology at UiT—The Arctic University of Norway, as well as being a publisher and filmmaker. He has been a main figure of the Nordic Anthropological Film Association (NAFA) since the late 1970s and has written extensively on visual anthropology and ethnographic filmmaking in key publications such as *Film as Ethnography* (1992, ed. with David Turton) and *Reflecting Visual Ethnography* (2006, ed. with Metje Postma).

Greg de St. Maurice is a postdoctoral research fellow at the University of Toronto and l'École des Hautes Études en Sciences Sociales. An anthropologist whose research focuses on place, taste, globalization, foodways, and Japan, he serves as Vice President of the Association for the Study of Food and Society. Recent publications include

"Kyoto Cuisine Gone Global" (in *Gastronomica*) and "Everything but the Taste: Celebrating Kyoto's Shishigatani Squash as Culinary Heritage" (in *Food, Culture, and Society*).

Amanda S. Green is an applied cultural anthropologist (Ph.D. Oregon State University) who specializes in Indigenous food movements of the Arctic with a focus on Sámi food production and food activism in Sweden. She served as a postdoctoral fellow in Environmental Studies at Davidson College and is an Assistant Professor of Anthropology at Eastern Kentucky University. Her research on food activism has been published in *Practicing Anthropology*.

Joan Gross is Professor of Anthropology at Oregon State University where she helped start the Food in Culture and Social Justice program. She is author of *Speaking in Other Voices: An Ethnography of Walloon Puppet Theaters* (2001) and co-author of *Teaching Oregon Native Languages* (2007). Her current research focuses on food activism in Oregon and Ecuador. She has chapters in *Food, Agriculture and Social Change: The Vitality of Latin America* (2017) and *Food Health: Nutrition, Technology, and Public Health* (2017).

Line Hillersdal is Assistant Professor in the Department of Anthropology, University of Copenhagen. Her work centers on the body in biomedical context, life with disease, and interdisciplinary collaboration. Key publications include: "Changing tastes: learning hunger and fullness after weight-loss surgery" (with Bodil Christensen and Lotte Holm, 2016).

Susanne Højlund is Associate Professor in the Department of Anthropology at Aarhus University in Denmark. She has done research on children, childhood and welfare, home, food, and taste anthropology in Denmark and Cuba. She has published widely on methodology and ethnography with children, has published a Danish book on taste, *Smag* (2016), and is co-editor of *Sugar and Modernity in Latin America: Interdisciplinary Perspectives* (2014).

Katy Overstreet is completing a joint PhD in anthropology at University of California Santa Cruz and Aarhus University. She works with the Aarhus University Research on the Anthropocene (AURA) project where she has been involved in transdisciplinary research about the environmental challenges facing the planet. Her research interests include more-than-human relations, bioindustrialization in agriculture, infrastructures of (re)production, and more-than-human embodied knowledges.

Jens Sejer Østergaard Rasmussen holds a masters in visual anthropology from Aarhus University. He specializes in anthropological filmmaking and qualitative market research. He is interested in taste as a social sense, intersensoriality, and group-making. More specifically his academic work examines how food is tied into social relations and how the taste of food can act as an empirical window into the formation of knowledge and communities.

Christy Shields-Argelès is a cultural anthropologist, Assistant Professor of Global Communications at The American University of Paris, and an associate of the Interdisciplinary Institute of Contemporary Anthropology in Paris. She has studied the cross-national comparison of dietary discourses and alimentary identity constructions in France and the United States as well as the culinary narratives of the American wives of Frenchman. Her current research focuses on the tasting practices of Comté cheese producers in eastern France.

Valeria Siniscalchi is Associate Professor of Anthropology at the École des Hautes Études en Sciences Sociales (EHESS), Centre Norbert Elias, Marseille, France. Her work focuses on economic anthropology, food activism, and the relationship between food, social movements and politics in Italy, the French Alps and the Slow Food movement. She is co-editor of *Food Activism: Agency, Democracy and Economy* (2014). She is completing a monograph *Slow Food. The Economy and Politics of a Global Movement.*

David Sutton is Professor of Anthropology at Southern Illinois University Carbondale. Since the early 1990s he has been conducting research on the island of Kalymnos and has published two books on the food culture of the island: *Remembrance of Repasts: An Anthropology of Food and Memory* (2001), and *Secrets from the Greek Kitchen: Cooking, Skill and Everyday Life on an Aegean Island* (2014), exploring questions of sensory experience and everyday life.

Amy B. Trubek is Associate Professor and Faculty Director of the Graduate Program in Food Systems at the University of Vermont. Her research interests include cooking as a cultural practice, taste of place, globalization of food systems, and culinary and sensory education. She is the author of *Haute Cuisine: How the French Invented the Culinary Profession* (2000), *The Taste of Place: A Cultural Journey into Terroir* (2008) and *Making Modern Meals: How Americans Cook Today* (2017).

Penny Van Esterik is Professor Emerita of Anthropology at York University, Toronto. Her fieldwork was primarily in Southeast Asia. She is a founding member of WABA (World Alliance for Breastfeeding Action). Past publications include *From Virtue to Vice: Negotiating Anorexia*, *The Dance of Nurture: Negotiating Infant Feeding* (both with Richard O'Connor), *Beyond the Breast-Bottle Controversy*, and *Materializing Thailand.*

Preface

Carole Counihan and Susanne Højlund

We began our collaboration on food anthropology and taste via email in 2013 and continued in 2014 when Susanne invited Carole to Aarhus University to be visiting professor for a month. We decided to organize a panel at the 2015 American Anthropological Association Annual meeting in Denver, Colorado, on "Making Taste Public." The enthusiastic response to our call for papers and the high quality of the presentations affirmed our commitment to forge on. Two years later we are nearly there, and very much looking forward to sending our common "child" out in the world.

We thank everyone who participated along the way. At Aarhus University, we thank the Department of Anthropology, the Faculty of Arts, and the AUFF fund for sponsoring travels, hosting meetings, and making possible a writing retreat to the wonderful Klitgaarden in September 2017 to pull the book together. We would also like to thank the Taste for Life colleagues and the Nordea Foundation for both intellectual and economic support. We especially thank the contributors to this volume for their fascinating work and their patience with our requests for revisions. At Bloomsbury, we thank Miriam Cantwell, Lucy Carroll, and the anonymous reviewers for their feedback and good ideas. As always, Carole thanks her husband Jim Taggart for years of steadfast support and intellectual companionship. Susanne thanks her faithful supporter Kurt and her family for many good discussions about the importance of taste.

Carole: thanks to Susanne for the invitation to Aarhus University in May 2014 to teach about food and culture and learn about your research on taste, and that of your students. Thank you for an especially warm introduction to Denmark, its academic life and its people, and most importantly to your wonderful family and husband Kurt.

Susanne: I am grateful to Carole that you took the chance to visit Denmark, that you believed in the idea, and that you have generously shared your knowledge and experience as a food anthropologist and an editor. A warm thanks to Jim for having engaged so enthusiastically in our work, and to you both for welcoming us into your home!

Foreword

David Sutton

Refugees and asylum seekers caught in long-term "hot spots" in Greece and Italy face many challenges, not the least of which is certainly the taste of food. Numerous protests have been launched over inedible or simply crushingly tasteless rations such as the plain pasta that makes up the majority of rations of asylum seekers in Macerata province in Italy. Or as one Libyan man who has been in Italy for nine months awaiting word on his asylum application put it, "Pasta—Bianco—Pasta—Bianco—Pasta—Bianco, every day only pasta," decrying with verbal repetition the numbing repetition of plain pasta. "Basta [enough] Pasta" he quips.[1] This simple request for variety and flavor strikes at the all too common sense that people in desperate straits cannot be "choosers," and that in these circumstances nutrition is the only thing that matters. The fact that the issue of flavor has been one (among many) flashpoint, however, is a reminder that *taste matters*, and not just to the much derided "coastal elites" of Wiliamsburg, Brooklyn or the Bay Area.

Also, "deliciousness cannot be the entirety of the narrative." So writes Tunde Wey, chef and writer, reflecting on the mundane and widely accepted aspects of white supremacy in the wake of the Charlottesville protests of August 2017. Wey explores the domination of "Best Restaurant" lists by white establishments, including the downscale sandwich shop *Turkey and the Wolf* in New Orleans, whose many African American rivals don't ever seem to rate a mention: "The food was delicious, but most of the memory of it faded after the last bite."[2] As he concludes, "Let food forever be the metaphor, and food stories the allegory. . ."—a cogent reminder of the public, political nature of taste.

The chapters in this collection all in different ways take us into the public, and often political nature of tasting. Over forty years ago, Clifford Geertz (1973: 12) famously wrote "Culture is public because meaning is." This constituted his shot across the bow at an anthropology that might have previously neglected meaning because it was internal to peoples' minds and thus not accessible to the anthropologist. While few now doubt this point, taste has struggled with a similar fate: how can we write about taste when it happens on tongues that we cannot observe? Or even worse, if it is seen by cognitive neuroscientists to happen not on tongues but to be locked in individual minds, accessible only to the laboratory elicitation techniques that either dispense with subjectivity or simply quantify it (Howes 2016).

Making taste public means following taste outside of brains and bodies (which play a part, of course) into the messy world of social and sensual practices that anthropologists are familiar with. Where does taste happen? No doubt it happens in the conduits between our senses and the world. Given the rise of food studies in anthropology over the past fifteen years it is surprising to say this, but this collection is the first sustained attempt to explore what tasting in public might mean for ethnographic exploration.

Taste, as phenomenon, or tasting, as practice, is a boundary object, very much "caught in between" (Højlund) traditional distinctions such as subjectivity and objectivity, individual choice and cultural determination, and deeply embedded in contexts of various kinds amenable to anthropological exploration and delineation. It is this aspect that makes this collection not just ethnographically, but also theoretically cutting edge as well, as we explore taste that happens in the interstices among people, people and objects, people and animals, and people and the materiality of food. Taste happens not just in tongues and brains, but in the spaces that connect us: in between mother and in-utero fetus (Van Esterick), between human and non-human eaters calling for "recursive forms of attunement" (Overstreet), entwining chefs and their tools (Østergaard), and between health professionals, their patients and the patients' families (Christensen and Hillersdal). Tastes are malleable; they take on multiple influences as they form and reform in-between school classmates (Højlund), through the "sensory attunement" of participants in tasting panels (Counihan), or are negotiated collectively among wine cooperative members drawing on memory and cultural continuity (Black). Tasting can be thought of as a skill or education of attention (Trubek and Carabello), a moral imperative (Siniscalchi), or a ritual process of returning to the whole (Shields). Tastes become public also in the context of macro-factors such as cultural contact and penetration (Crawford), struggles for food sovereignty in the context of climate change (Green), changing neoliberal trade policies (Gross), as well as attempts to counter these decontextualizing and "disembedding" trends (Shields). Social contexts, then, are holistic, encompassing "mind, body, people, places, and comestibles" (Counihan and Højlund), but so is, of course, the sense of taste itself, relying as it does on far more than the tongue (and nose), but rather a synesthetic, or multisensory engagement (Gross, St. Maurice). You can "listen to smells" as Seremetakis (1994) noted for Greek grandmothers, food essentially starts by "looking good" (or bad) as Philosopher Carolyn Korsmeyer (2011) argues. It can happen "in the hands" (Mann et al. 2011), or in the stomach (Mann 2015). Talk of attunement of the senses takes me back to the work of my mentor James Fernandez (1986), who way back in the 1980s when the senses were far from our anthropological agenda was writing of the "mutual tuning in" that is an essential part of creating the con-sensus, or coming together of the senses that was as important for making metaphor as it was for rituals of "revitalization" in Gabon as for "conviviality" in Spain. All of these word choices suggest that sensibility and sociality are ineluctably tied together, just as sensing is with making sense (Howes 2016). And it is a reminder not only of the social production of the sensory, but of the sensory production of the social as well (see Chau 2008; Sutton 2016).

Capturing taste, then, means resisting the urge to pin it down, to specify its locations and limits. Rather, we should indeed make taste public by opening it to the world and following where it leads us into new contexts and conjunctures. This collection does exactly that.

Notes

1 See http://theconversation.com/enough-pasta-already-why-asylum-seekers-in-italy-are-fed-up-with-their-food-rations-84147.

2 See www.sfchronicle.com/restaurants/article/Look-to-the-food-world-to-understand-America-s-11950975.php.

References

Chau, A.Y. (2008). "The Sensorial Production of the Social." *Ethnos* 73: 485–504.

Fernandez, J. (1986). *Persuasions and Performance: The Play of Tropes in Culture.* Bloomington: Indiana University Press.

Geertz, C. (1973). *The Interpretation of Cultures.* New York: Basic.

Howes, D. (2016). "The Science of Sensory Evaluation: An Ethnographic Critique." In *The Social Life of Materials*, A. Drazin and S. Kuchler eds., 81–97. New York: Bloomsbury.

Korsmeyer, C. (2011). "Looks Good!" *Food Culture and Society* 14(4): 462–468.

Mann, A. et al. (2011). "Mixing Methods, Tasting Fingers." *Hau* 1(1): 221–243.

Mann, A. (2015). Tasting in Mundane Practices: Ethnographic Interventions in Social Science Theory. University of Amsterdam: Digital Academic Repository.

Sutton, D. (2016). "'Let them Eat Stuffed Peppers': An Argument of Images on the Role of Food in Understanding Neoliberal Austerity in Greece." *Gastronomica* 16(4): 8–17.

1

Making Taste Public

An Ethnographic Approach

Carole Counihan and Susanne Højlund

Introduction

Anthropologists have shown that different cultures have different taste preferences and flavor principles embedded in cuisine. But how does taste become part of culture? Taste is often described either as a characteristic of food, or as an individual perception of or response to an outer (cultural) world. Instead of seeing our senses as passive receptors, impacted or shaped by culture, we turn the perspective around and ask how we use our senses to bring taste from a private experience to be part of the public. We define "the public" as the social space between people and food, a space where taste becomes accessible for others to engage in and possibly to share. We assume that human interaction, communication, and practices of distribution shape taste, and the central focus of this book is how this happens. We take up David Sutton's (2010: 220) call to anthropologists to examine, "everyday life and the multiple contexts in which the culturally shaped sensory properties and sensory experiences of food are invested with meaning, emotion, memory, and value."

In this book anthropologists use ethnographic case studies to explore how people share our senses of taste and the experience of eating, with whom, and under what conditions. The chapters focus on how the taste of food—the sensual experience and the preferences, identities, and meanings associated with it—is and becomes social. We think about how tastes are made and shared between us in public settings where they are sometimes exalted and sometimes abhorred. We examine how tastes are socialized, who are the key actors, what are the key sites, where are the spaces of taste exchanges and clashes, and how values are attached to tastes. We analyze how through different forms of exchange tastes become part of our sensorial apparatus and identity in a range of cultural contexts.

Chapters focus on ethnographic examples combined with theoretical discussions of the process of making taste public. They examine definitions and mobilizations of taste in different sites, institutions, public places, and regions around the world, depicting ethnographic understandings of how people learn and practice taste. Cases include the construction and sharing of tastes in infant feeding (Van Esterik) and Danish and Solomon Island meals (Christensen and Hillersdal, Crawford), cooking lessons in Danish and North American schools (Højlund, Trubek and Carabello), Danish and Japanese chefs' constructions of meals (Østergaard, de St. Maurice), French cheese tasting panels (Shields), Wisconsin dairy farmers' choice of feed for their cattle (Overstreet), northern Italian viticulturalists' production practices (Black), and food activists' political strategies in Italy and Sweden (Counihan, Siniscalchi, Green).

Why Study Taste?

There is a huge tradition in anthropology on the study of food, so why should we need to outline one more food related field? What makes taste more significant to study than "food"? First, the concept of taste expresses *a relation* between humans and their food. When taste is put in the forefront we are looking for definitions, interpretations and experiences of this relationship. Without denying the physiological dimensions (Shepherd 2012; Spence and Fiszman 2014), we focus on taste as a social construction and especially on how people are externalizing their sensory knowledge (cf. Berger and Luckmann 1991 [1966]). We are interested in how attractions and repulsions, pleasure and disgust, meanings and opinions are created in the space between people and their food and how they lead to different definitions of quality, and to shared social platforms for performance, lifestyle and identity. In taking this perspective we are very much in debt to the writings of Genevieve Teil and Antoine Hennion (2004), who have created an analytical approach for thinking about taste as activity, reflection, and performance. Thus, in choosing "taste" and "tasting" as analytical concepts we stress food and eating in specific social, sensory, and cultural contexts.

Second, we want to expand and challenge the argument that a special interest in taste is either a luxurious interest or driven by necessity (Bourdieu 2013). We acknowledge that taste also is about class and lifestyle distinctions, and that there are moments where nutrition and coping with hunger must stand at the forefront rather than taste. But this does not mean that only elites should care about taste, or that taste does not matter in situations of scarcity. We hope to show that taste is a central daily part of people's eating habits, affecting what and how much they consume, what it signifies, and whether it satisfies not just their nutritional but also their emotional and social needs. Following this argument, we see taste as a societal issue, e.g. playing a role in obesity epidemics (Schatzker 2016), being part of a larger aesthetic economy (Michalski 2015), and creating conflicts and debates on national identity (Tellström et al. 2003). Throughout the book, we thus argue for a broader perspective on taste involving analyses that understand people's everyday relationships to their food in a comprehensive social context.

Third, we find it important to contribute to the newer discussions on how we sensorialize our world (Chau 2008). Traditionally taste has been excluded from such analyses, giving primacy to sight and hearing (Korsmeyer 1999). These two faculties have been objects for the "agency paradigm," which conceptualizes them as active senses rather than passive. Scholars have discussed "ways of seeing" (Berger 2008), and "ways of listening" (Clarke 2005), but no-one has conceptualized "ways of tasting" although David Howes and Constance Classen (2013) come close in their book *Ways of Sensing*, where they also remind us that sensing is a synaesthetic activity (ibid:152). We contribute to these discussions by seeing taste as the product of an holistic engagement of all the senses with food mediated by the complex relationships between mind, body, people, places, and comestibles. We propose that analyses of taste focus on the multisensorial agency of people, and on how taste is produced, either on the plate, in our communication, through our hands and craftsmanship, or in our sharing of values and activities.

Even though we stress this "production side" of our sense of taste we recognize that culture creates a context that cannot be ignored. Taste is an irrevocable part of culture and therefore impacted by it and generated in relation to it. Joan Gross's chapter, for example, shows how Latin Americans' concepts of "fresh" are linked to their home place and to just-harvested local products, which are far from North American meanings of fresh. Crawford's chapter shows that taste may not emerge as a cultural concern until contrasts with new foods arise and call out taste distinctions, much as Rick Wilk (1999) found that in Belize the concept of local food only developed when global food arrived. Taste arouses passionate feelings in many, but there are cultures where people seem to have little investment in it, such as the Kenyan Samburu pastoralists (Holtzman 2009: 157) or the Solomon Islanders in Crawford's chapter. Taste can stimulate experimentation and learning about new foods and forge connections to new cultures and places, as in de St. Maurice's chapter on the mission of the Japanese Culinary Academy. But perceived bad tastes can also produce and signify distance, rupture, and exclusion, as Siniscalchi's analysis of Slow Food reveals (see also Rhys-Taylor 2016; Walmsley 2005).

Methods for Sensory Ethnography

There is growing interest in "sensuous scholarship" (Stoller 2010), "sensory anthropology" (Howes and Classen 2013: 11), and "sensory ethnography" (Pink 2009), but until recently there has been a dearth of clear direction on *how* to study the senses of people in other cultures, even in important texts like Pink (2009), Stoller (2010), and Sutton (2010). Dara Culhane and Denielle Elliott's (2016) *Different Kind of Ethnography*, is a step in the right direction and we accept their challenge to take sensory experience seriously as "epistemological and political critique" (11). Dara Culhane (2016) lays out how anthropologists can "begin paying close attention to sensory experience in a critical, purposeful way" with exercises to attend to, record, and present what we see, hear, smell, taste, and touch. She aims for "sensory embodied reflexivity" (61)—a deeper awareness of one's own and thus others' sense perceptions.

Several researchers have talked about using their own bodies to take in sensory experience and to develop critical awareness of how their own sensory biographies affect their perceptions (Chau 2008; Rhys-Taylor 2016; Stoller 2010). Here Penny Van Esterik describes her own overwhelming disgust at the taste of fish to hypothesize about how early-childhood feeding may set up life-long taste preferences. Rachel Black (2017) provides useful exercises to stimulate students' sensory acuity, e.g. having them compare different qualities of pizza dough and shape it totally by hand, and then adorn, bake, admire, smell, and finally taste it.

Being there with all the senses unfurled is critical to sensory ethnography (Black 2017; Culhane and Elliott 2016; Pink 2009). Anthropologists' longstanding tool of participant observation broadens to become "participant sensation" (Howes 2006:121)—here demonstrated particularly richly in Katy Overstreet's chapter "Taste Like a Cow." She describes accompanying a farmer as he walks through a hay sale and decides which hay to buy for his cows, joining him in visual assessment of the hay's texture and color, handling it to feel its smooth or "pokey" texture, and smelling it for freshness or mold. Højlund's chapter relies on participant-sensation in a different way as she uses all her senses to gather the diverse perceptions of Danish middleschoolers. She also records their words and interactions surrounding taste at school cooking classes and interviews, and participates in making food and evaluating the tastes produced.

We believe that recording how people talk about taste at interviews, tastings, meals, and other events is an important addition to participant-sensation and fieldnotes. Trubek and Carabello's chapter is based on recordings of cooking practices and subsequent interviews with students in a university Food and Culture course. Subjects' words are a vivid window into their subjective insider perceptions of taste. Videorecording is the most complete because it includes facial and tactile gestures, body language, and relationships, as well as words and tone of voice, but it is complex to carry out and analyze, whereas voice recording—accompanied by photos if possible—is simpler to carry off. Transcriptions are excellent for capturing feelings and meanings about taste, as Counihan's recording of a Sardinian caper tasting reveals. In interviews researchers can deepen discussion and elicit metaphors, memories, and comparisons by posing questions such as "What does it taste like? What does it remind you of? How does it compare to other things you have tasted?"

Ethnographers still face the challenge to find sensually rich ways to communicate beyond our trusty reliance on the written word. Culhane (2016) encourages dance, music, performance, fiction, photography, and film. A fine example of a multi-dimensional, multi-sensorial exploration of taste was the "Creative Tastebuds" symposium held at the Aarhus Theater in Aarhus, Denmark, September 4–5, 2017.[1] Four panels addressed "how brain and culture collaborate on taste," where creative mediators—a writer, chef, architect and performance artist—pushed four pairs of natural scientists and anthropologists to converse about taste. There were "innovation showcases" with tastings, sound-taste labs, art projects, and performances; "soap box dialogues" with the public about topics of interest; on-site lunches to experience taste through commensality; and a thoughtfully prepared, delicious, and convivial dinner for all conference attendees. Participants had the opportunity to engage socially, intellectually, sensually, emotionally, and corporeally with taste.

Taste Socialization: Family and Culture

An important theme of this book is that tastes are socialized through people's interactions with others—in the family and other social groups. Van Esterik explores humans' earliest feeding experiences and ponders how they might affect lifelong decisions to embrace or reject new tastes. The mother's foods and flavors pass to the fetus in utero and continue to pass to the infant through breastmilk, socializing the child to familial and cultural tastes—whether chili peppers, hotdogs, or camembert. Van Esterik asks about the taste repercussions of feeding infants industrially produced formula based on cow milk or soy protein. Formula may well predispose them towards "the industrial palate" of salty sweet artificial flavors that typify processed foods, whose production is the focus of recent work by Ella Butler (2017), Anna Mann (2018), and others.

Quotidian meals express changes in taste over forty years in Crawford's chapter on Solomon Islanders whose acceptance of canned fish and spicy foods calls forth the blandness of the traditional diet and offers an interesting case study of the widespread phenomenon of dietary transformation. Daily meals are the battle ground for changing family diets and taste preferences in Christensen and Hillersdal's chapter on women undergoing weight loss surgery in Denmark, whose need to transform their post-surgery eating habits has repercussions on family dynamics, gender roles, and commensality.

Taste, Place, and Intersubjectivity

A key determinant of taste is place—the places where plants and animals are raised and processed into comestibles, and the places where they are eaten. Amy Trubek (2008) has written about the taste of place or *terroir*—which refers not just to climate and landscape but also to the skills, traditions, and human and animal relationships involved in making food. In her chapter on "taste landscape," Black learns the taste of Carema wine by hiking through steep northern Italian vineyards, convivially drinking as well as formally tasting diverse vintages, and interviewing producers. She shows how the evolving relationship to place due to mechanization, trellising, cooperative production, and export markets changed the taste of Carema wine, but nonetheless its importance to local identity has persisted. Similarly, *terroir* is essential to the taste of Comté cheese, the focus of Shields's chapter. She focuses on how the *jury terroir*, a panel of trained volunteers, defines Comté cheese at their monthly meetings where they taste and describe diverse exemplars. This practice enhances jury members' feelings of relatedness to each other and the wider natural world, showing how taste is social and intersubjective. Katy Overstreet's chapter extends that taste intersubjectivity from humans to animals by exploring the many sensory ties between Wisconsin dairy farmers and their cows around selection of the best and tastiest feed.

Taste Education and Sharing: Identity and Community

Because of its ability to make connections, taste plays an important part in teaching and learning across a variety of occasions and institutions. De St. Maurice explores how a formal institution, the Japanese Culinary Academy (JCA), musters various strategies to make the taste of Japanese cuisine public, shared, and authentic to strengthen Japanese identity and traditions against the threat of globalization. The JCA project reveals an understanding of taste as a multi-sensory phenomenon inextricably linked to cultural values. The emphasis on all the senses also pervades the cooking classes at the University of Vermont studied by Trubek and Carabello and the Danish middle schools observed by Højlund. These encourage students to fully engage with the sight, smell, touch, taste, and sound of food. Moreover the classes emphasize the important social dimension of learning, and students develop their skills and sensory perception by making taste together.

Taste, however, acts not just as glue but also as disruptor. This is revealed in Østergaard's chapter on Danish professional chefs' interactions in a restaurant kitchen. They learn and share understandings of taste through their adherence to the kitchen's prevailing collective know-how that develops skills, practices and relationships to create distinctive tastes from raw materials. But an apprentice lost his job because he failed to adapt to the material and social practices of the kitchen and produce the collectively sanctioned tastes.

Taste Politics

Taste is often implicated in power relations, hierarchy, inclusion, and exclusion. Bourdieu (2013: 31) has famously noted the "basic opposition between the tastes of luxury and the tastes of necessity," referring to how elites can play with tastes but common people must eat what they can get. Even in conditions of scarcity, however, people care about taste, and its recollection may be a means to transport them to better times, as happened when the Jewish women starving in the Terezin concentration camp wrote down and exchanged recipes of favorite dishes they could only savor in their minds (De Silva 2006).

Taste has become instrumental in diverse activists' efforts to transform the food system and establish food sovereignty such as those of the Sámi reindeer herders of northern Sweden discussed here by Amanda Green. Reindeer fat from free-range animals has a distinct taste of place from their diet of lichens, mushrooms, and grasses. Activists educate consumers to distinguish this particular taste and understand its dependence on access to land and Sámi cultural practices—both threatened by encroaching state power and food globalization. Eating the free-range reindeer fat enacts resistance to those threats.

In Italy, taste has longstanding salience and activists use it as a lever to promote local and sustainable food and agriculture. Counihan's chapter discusses Slow Food's efforts to support Sardinian capers by exalting their tastes and teaching about them at a commensal caper tasting. This case supports Sarah Pink's (2009: 73) affirmation that you can learn a lot about taste while people are eating together. Siniscalchi's chapter examines Slow Food's construction of taste as a moral, economic, and political tool of the movement. Slow Food creates boundaries between "bad" and tasteless industrial food, shoddy producers, and non-discerning eaters on the one hand; and "good" local producers, their products, and the consumers who recognize their quality on the other. Taste is a powerful lever to manipulate consumption, resist globalization, and preserve heritage cuisine and national identity.

Lingering Questions

This collection of ethnographic essays on making taste public hopes to spur further research. We need studies that extend what we know about artificial flavors and how they affect our relationship to food, plants, and animals. We need to know more about institutions that feed lots of people—schools, hospitals, prisons, and workplace cafeterias—and how they mobilize, diversify, or ignore taste. We need to continue to explore how diverse and unfamiliar tastes can be bridges rather than barriers between people across race, ethnicity and national origin, and we need to explore how gender affects taste perception and socialization across class and ethnic lines. We need to know how families teach taste and inculcate or disrupt preferences across generations. And last, we need to know how we can use taste to bring just, sustainable, healthy, and delicious food to all.

Notes

1 "Creative Tastebuds" was led by Susanne Højlund and Mikael Schneider and sponsored by the European Capital of Culture Aarhus 2017 and the Aarhus/Central Denmark Region 2017—European Region of Gastronomy, Taste for Life/Smag for Livet (which is funded by Nordea-fonden), University of Aarhus, More Creative/Region Midtjylland and Fonden Aarhus 2017. See http://creativetastebuds.dk accessed October 4, 2017. Conference papers are forthcoming in a special issue of the *International Journal of Food Design* edited by Ole G. Mouritsen and Michael Bom Frøst.

References

Berger, J. (2008 [1972]). *Ways of Seeing*. London: Penguin Books.

Berger, J. and T. Luckmann. (1991 [1966]). *The Social Construction of Reality. A Treatise in the Sociology of Knowledge*. New York: Anchor Books.

Black, R.E. (2017). "Sensory Ethnography: Methods and Research Design for Food Studies Research," in J. Chrzan and J. Brett, eds., *Food Culture: Anthropology, Linguistics, and Food Studies, Research Methods for the Study of Food and Nutrition* vol. 2, 228–238. New York: Berghahn.

Bourdieu, P. (2013 [1979]). "Distinction: A Social Critique of the Judgment of Taste," in C. Counihan and P. Van Esterik, eds., *Food and Culture: A Reader*, 3rd ed., 31–39. New York: Routledge.

Butler, E. (2017). "How to Talk about Taste?" Paper delivered at the Stop Making Sense Conference. Philadelphia, PA, March 10, 2017.

Chau, A.Y. (2008). "The Sensorial Production of the Social." *Ethnos: Journal of Anthropology* 73(4): 485–504.

Clarke, E.F. (2005). *Ways of Listening: An Ecological Approach to the Perception of Musical Meaning*. Oxford: Oxford University Press.

Culhane, D. (2016). "Sensing," in D. Culhane and D. Elliott, eds., *A Different Kind of Ethnography: Imaginative Practices and Creative Methodologies*, chapter 3. Toronto: University of Toronto Press.

Culhane, D. and D. Elliott, eds. (2016). *A Different Kind of Ethnography: Imaginative Practices and Creative Methodologies*. Toronto: University of Toronto Press.
De Silva, C., ed. (2006). *In Memory's Kitchen: A Legacy from the Women of Terezin*. Lanham, MD: Jason Aronson.
Holtzman, J. (2009). *Uncertain Tastes Memory, Ambivalence, and the Politics of Eating in Samburu, Northern Kenya*. Berkeley: University of California Press.
Howes, D. (2006). "Charting the Sensorial Revolution." *Senses and Society* 1(1): 113–128.
Howes, D. and C. Classen. (2013). *Ways of Sensing*. Florence, GB: Routledge.
Korsmeyer, C. (1999). *Making Sense of Taste. Food and Philosophy*. New York: Cornell University Press.
Mann, A. (2018). "The Scientization of Taste." *Food and Foodways* 26, (1): forthcoming.
Michalski, D. (2015). *The Dialectics of Taste: On the Rise and Fall of Tuscanization and Other Crises in the Aesthetic Economy*. New York: Palgrave Macmillan.
Pink, S. (2009). *Doing Sensory Ethnography*. London: Sage.
Rhys-Taylor, A. (2016). *Food and Multiculture: A Sensory Ethnography of East London*. Oxford: Bloomsbury Academic.
Schatzker, M. (2016). *The Dorito Effect: The Surprising New Truth About Food and Flavor*. New York: Simon & Schuster.
Shepherd, G. (2012). *Neurogastronomy*. New York: Columbia University Press.
Spence, C. and B.P. Fiszman. (2014). *The Perfect Meal: The Multisensory Science of Food and Dining*. Oxford: Wiley Blackwell.
Stoller, P. (2010). *Contemporary Ethnography: Sensuous Scholarship*, Philadelphia: University of Pennsylvania Press.
Sutton, D.E. (2010). "Food and the Senses." *Annual Review of Anthropology*, 39: 209–223.
Teil, G. and A. Hennion. (2004). "Discovering Quality of Performing Taste: A Sociology of the Amateur," in M. Harvey, A. McMeekin and A. Warde, eds., *Qualities of Food*, 19–37. Manchester: University of Manchester Press.
Tellström, R., I. Gustafsson and C. Fjellström. (2003). "Food Culture as a Political Tool—Meal Construction during the Swedish EU-chairmanship 2001." *Food Service Technology*, 3(2): 89–96.
Trubek, A.B. (2008). *The Taste of Place: A Cultural Journey into Terroir*. Berkeley: University of California Press.
Walmsley, E. (2005). "Race, Place and Taste: Making Identities Through Sensory Experience in Ecuador." *Etnofoor*, 18(1): 43–60.
Wilk, R. (1999). "'Real Belizean Food': Building Local Identity in the Transnational Caribbean." *American Anthropologist* 101(2): 244–255.

Part One

Taste Socialization: Family and Culture

2

Taste, Socialization, and Infancy

Penny Van Esterik

Introduction

The subject of infant feeding, particularly breastfeeding, is fraught with tensions, both personal and theoretical. This chapter explores how taste is developed over an individual's lifetime and is maintained or transformed across generations. What if decisions to embrace new tastes or refuse new tastes are conditioned by initial feeding experiences? What if taste becomes social through the early developmental experiences of infant feeding? What if the taste of human milk, cow's milk or soy milk sets the template for appreciating the taste of future foods throughout life, and more important, establishes a trajectory that shapes the nature of social relations around eating? What would constitute evidence for this argument?

Breastfeeding and infant feeding provide a unique interactive space of taste socialization for exploration. The subject is most extensively researched by biochemical scientists and biopsychologists. Their lab logic has produced well-developed literatures documenting experimental evidence about the development of taste perception in infancy and childhood. Shepherd's work on *Neurogastronomy* (2012) also combines the biochemical and the sensory, but not the social and cultural. Early psychoanalytic work by Anna Freud (1946) and Melanie Klein (1928) stressed the long-term effects of weaning dilemmas and how the conflicts of the oral phase manifested in adult problems. The purpose of some of this work was to define the origin and meaning of disordered eating. Psychoanalysts present eating as a struggle between mother and child for control. Children, it seems, have most pleasure when they eat what they like, without being controlled by their mothers or primary caretakers.

A good scientist must control for or exclude from consideration confounding variables; confounding variables, better known as context, is where a good anthropologist begins. Lab logic of the sensory sciences raises many interesting questions for anthropologists to pursue. But there are other logics that also shed light on the subject of early feeding experiences. The field logic of ethnography moves beyond the level of individual preference and choice to consider the broader cultural context of taste socialization. But ethnographic opportunities to examine infant feeding in the field are limited because infants cannot be part of experiments that manipulate their infant feeding regimes, nor can they talk or text about their taste preferences.

Observational information is available from psychological work on taste perception and biochemical research on flavor transfer. We know that babies taste everything, that tasting as an activity is important to them. But we do not know what they taste, what things taste like to them. This makes the study of taste and infancy particularly challenging.

The Meaning of Taste

Taste in this volume emphasizes the gustatory rather than the artistic. Taste as the physiological response to our olfactory and gustatory senses and the cultural and social refinement of taste to assert difference are intimately related (Crowther 2013: 284). Bourdieu brings the two approaches to taste together, arguing that intuitive judging of aesthetic values is inseparable from the capacity to discern flavors in food, debunking the distinction between aesthetic and non-aesthetic senses (Korsmeyer 1999: 64). Bourdieu (1984) uses habitus to refer to the way the individual internalizes and embodies patterns of taste and distinction, mediating perception and sensory input. Taste as disguised class hegemony is often expressed through food preferences. Bourdieu uses the contrast between the taste of luxury and the taste of necessity to express the mix of the gustatory, the political and the moral dimensions of good taste. Taste as sense and aesthetic disposition is both natural and social. Taste standards develop out of a collective habitus that precedes aesthetic judgement. But infants do not yet know the public meaning of taste, only the immediate sensation of pleasure.

Where taste perception is hard wired in humans, taste preferences are related to individual experiences and cultural expectations. Textures and smells are the key to flavors and taste, and figure largely in food preferences. As Wilson explains, flavors are memories generated backwards through the nose (2015: 43). For each taste, we create a memory bank of flavor images that can last a lifetime (Wilson 2015: 48). How far back do these taste memories go?

Our memories of past tastes can be complicated by the differences between perceived and actual changes in the taste of different food items. For example, to me, grapefruits I eat today are sweeter than the ones I remember eating in the past, and tomatoes today taste completely different from those I remember. Canned peaches taste totally unlike fresh peaches. Seremetakis recalls similar changes in the taste of Greek peaches (2005: 297). Are these faulty flavor image memories, changing public perceptions about the taste of foods or have the fruits really changed in flavor over the past few decades? Food technology might hold some of the answers to the development of tasteless fruit, genetically modified to travel well and look good rather than taste good.

In adults, taste receptors vary from person to person, with some people being very sensitive to tastes such as sweet and sour, and others exhibiting quite dull taste perception (Korsmeyer 1999: 87). Some ability to taste is under genetic control, but taste preferences are mostly based on experience. It is difficult to educate the palate and the digestive system after food and taste preferences have become established

(Korsmeyer 1999: 93), although wine tasters and tea tasters, for example, can be trained to recognize subtle taste differences. With the decline of traditional cuisines and food customs, people have to be taught to perceive taste differences. This is one aim of the slow food movement and the taste education classes in European and Scandinavian schools that teach children the pleasure of the table (Wilson 2015: 244).

The Taste of Infancy

The editors ask how taste becomes public. Part of the answer to the socialization of taste and taste preferences may be found in infancy, as infants build their unique constitutions and begin to internalize the habitus of taste (cf. Van Esterik and O'Connor 2017). The point of departure is the taste transfers begun *in utero* between mother and fetus. These taste transfers either continue through breastfeeding or terminate at birth with the introduction of breastmilk substitutes based on cow's milk or soy milk. An infant's constitution then is either built on taste continuities or taste disjunctions. Unlike the diversity of food practices of adults, infants survive on a single food, milk. Let us consider this one food contrast—between human and artificial milk—to explore the argument that taste perception is shaped by infant feeding practices.

Newborn infants are not born senseless; a fetus can taste by thirteen weeks (Wilson 2015: 43). Should we think of that infant as an epicure with refined taste, or as a *tabula rasa* with no capacity to make taste distinctions? All newborns react with pleasure to the taste of sweetness, neutrally or positively towards saltiness, and with obvious distress to bitter tastes (Harris and Coulthard 2016: 114; Korsmeyer 1999: 88). Glutamate, the most abundant amino acid in human milk, is also present in the amniotic fluid, and provides the familiar umami taste appeal of breastmilk. It is often assumed that infants do not have the capacity to make taste distinctions and can only consume bland, flavorless foods because of their sensitive palates. The absence of strong taste defines food that is considered suitable for infants and young children. As a result, some societies may decide not to waste good food on children on the assumption that their tastes have not yet fully developed (cf. Dettwyler 1994).

Wilson writes: "it is depressing to learn that a person's food likes at age two generally predict their tastes at twenty" (2015: 42). Food memory is there at the start. Babies have nostalgia for tastes and are predisposed to respond to certain foods by their experience *in utero* (Wilson 2015: 43); they respond specifically to the odor of strong foods such as garlic and anise (Harris and Coulthard 2016: 114). Newborns arrive in the world with memories of how their amniotic fluid tasted and smelled (Wilson 2015: 44).

The first six months of life provide a sensitive learning period for the flavor image (Shepherd 2012: 235), creating a distinct and important sensory world adults know little about. As food passes into the pregnant woman's body, the nutrients also nourish the fetus through the placenta, the umbilical cord and the amniotic fluid. Food that is external to the body becomes internal, and is incorporated into the body—first by the mother, and then by the fetus, and finally by the breastfed child. Flavors from the pregnant mother's diet enter the amniotic fluid and are swallowed by the fetus throughout the pregnancy. Thus, the types of food, spices and drinks that pregnant women consume, guided by the flavor principles of their cuisines, are experienced by their babies long before their exposure through breastfeeding or eating complementary foods.

We have learned about this, not through social scientists, but through the lab logic of sensory scientists. Experimental evidence has shown that the flavors from foods ingested by pregnant women, including garlic, mint, vanilla, carrot, anise, and alcohol are present in amniotic fluid and breastmilk. Infants whose mothers consumed anise when pregnant showed a preference for anise (Shepherd 2012: 234). Hausner and her colleagues (2009) found that breastfeeding facilitates acceptance of novel flavors such as caraway. Similarly, in a study where pregnant women drink water or carrot juice in late pregnancy, the children of moms who drank water had more negative reactions to carrots than children of women who drank carrot juice. Although infants take pleasure in foods like carrots that they have tasted before, this enjoyment may not necessarily result in a greater intake of those foods (Mennella, Jagnow, and Beauchamp 2001). There is extraordinary variation in the rate of transfer of flavor compounds into breastmilk, as well as variation between women and even variation in individual women. Ingested flavor compounds are eliminated from mother's body quickly, but enter mother's milk selectively and in relatively low amounts, resulting in "continuous flavor changes in mother's milk" (Hausner et al. 2009: 123). These experiments show that taste memories from pregnancy and breastfeeding could provide the foundation for cultural differences in cuisine (Mennella et al. 2009: 780s). Mennella's research has shown that

> early flavor experiences gained through breastfeeding enhance the acceptance of a wider range of foods during the weaning period and throughout childhood. It has been argued that breastfed babies, whose mothers regularly eat a wide variety of foods, are exposed to a diversity of flavors that are absent from formula milk and this early exposure augments the acceptance of various flavors.
>
> Burnier et al. 2011: 200

Breastfed infants adapt to a wider range of foods than infants fed with commercial baby milks, and they are also more interested in trying new tastes; lab logic provides evidence that breastfed infants accept a wide variety of tastes and the idea of taste change itself, rather than simply acceptance of specific foods (Harris and Coulthard 2016). A longer duration of breastfeeding is associated with greater intake of fruits and vegetables and a lower intake of sugary drinks in six-year-olds (Perrine et al. 2014). Making your own baby food rather than using commercial products also predicts a greater consumption of fruits and vegetables (Harris and Coulthard 2016).

Sensory scientists tell us that breastmilk is predominately sweet and contains volatile food odors that vary from mother to mother. At birth, a newborn is attracted to the smell of its mother, and the smell and taste of human milk in the form of colostrum. While we are knowledgeable about the nutritional quality of human milk, we know less about its sensory qualities and taste.

Infant Formula and Taste

In spite of reports about ice cream and cheese made from human milk, it is easier to sample a bottle of infant formula than breastmilk, as infant formula is much like other commercial drinks served in bottles, like yogurt drinks or smoothies. Yet it is uncommon for adults to taste the milk-based products they prepare for their infants. An exception to this generalization was found on a blog (daddytypes.com 2004) where fathers of infants who were having difficulties digesting commercial formulas decided to perform a taste test on three popular brands. In brief, they found the taste and smell of the products unpleasant, to say the least. Soy-based formula and hypo-allergenic formula in particular were described as having an unpleasant, fishy or metallic aftertaste, gritty texture, and "smelled and tasted like poo." Parents on the blogs suggested masking the taste with sugar syrup or fruit juice, or adding breastmilk.

Although parents cannot know how their children were experiencing the taste of these products, one parent claimed their child noticed the taste difference between powdered infant formula and the ready to feed version of the same brand. In fact, the compositions of powdered and liquid formulas are often quite different. They may contain different sugars and fats; it would be interesting to know more about how infants become attuned to contrasts between different brands, and to different sugars and fats.

Regulations concerning breastmilk substitutes focus primarily on the promotional practices of formula companies and food safety; they seldom extend to the taste or how they meet (or fail to meet) the nutritional needs of infants. In fact, formula is one of the most basic substitute food products developed by the food industry. Like all industrial food products, they can be adulterated or contaminated. Baby food manufacturers add chemically created or genetically modified constituents such as pre- and probiotics as ingredients in infant formula. The outcome measure is often infant growth, with the overall message that bigger is better. This is one way to link use of infant formula to over feeding and later childhood obesity.

Formula-fed babies who delay getting solids until six months become children with limited tastes, and a taste preference for foods that may set them up for a lifetime of unhealthy eating (Wilson 2015: 25). Wilson speculates that infant formula, because its flavor does not vary, has more "imprinting" power than breastmilk (2015: 45). Building on this knowledge of imprinting, formula companies began adding vanillin to toddler milks, and a few added it illegally to infant formula. Once imprinted on the vanillin flavor, children are easily attracted to the taste of commercial cookies and cakes; everything tastes better with a vanilla flavor added. The addition of vanillin even appeals to adults who taste the milk from a baby bottle and like it (Wilson 2015: 26, 7). Mosby points out that changes in the regulations guiding the composition of baby foods such as banning MSG are made "to please the taste of mothers" (2009: 209, cf. Bentley 2014: 146).

Manufacturers are aware that special hydrolyzed formula for infants with allergies to milk "tastes and smells horrible." The more a formula is hydrolyzed to break down the protein, the worse it tastes. Some experimental nutrition studies usually funded through the baby food industry have explored the hedonic tone—whether the taste was pleasurable—of hydrolyzed protein hydrolysate formulas (ePHF) compared to cow's milk based formula. The former has pronounced bitter, sour, and savory tastes. Infants have heightened sensitivity to flavors around four months of age, and the experiment suggested that at least three months' duration of exposure to the taste of infant formula is necessary for establishing flavor preferences (Mennella and Castor 2012). It is not clear how long the potential effects of taste programming of infant formula persist. In order to explore this, infants were fed with hydrolysate formula which the researchers confirmed has a nasty sour cheesy tang with a hay-like odor. They found that infants fed on one of these nasty tasting formulas developed a preference for that taste. Bitter hydrolysate formulas are better accepted if they are started shortly after birth. Babies on these formulas learned that was how foods should taste. A study about how infants liked the taste of baby foods excluded infants who had consumed hydrolysate formula on the assumption that the formula would alter their ability to taste (Medrelle et al. 2017: 275). Even at four or five, children still preferred these sour, bitter tastes (Wilson 2015: 25; Harris and Coulthard 2016: 114). One study showed that the exposure to the tastes of these formulas as infants influenced the acceptance of infant formula among German ten-year-olds (Sausenthaler et al. 2010).

If infants cannot tolerate milk-based infant formulas, they are often prescribed a soy-based product. Soy milk products, developed in the 1920s, expanded rapidly in the fifties to include a number of infant formulas. Although widely recommended by pediatricians for infants who could not tolerate milk-based formula, some doctors warned that infants who cannot tolerate milk are also unlikely to tolerate soy products. Soy milks may contain as much protein as cow's milk, but the processing removes most of the calcium. Like cow's milk formula, the soy products are usually fortified. Vegans and others seek out plant-based milks made from soy, almond, cashew, or coconut, although there is no precedent in nature for feeding young mammals a plant-based protein (Sears 1999).

Around 25 percent of American babies consume soy-based infant formula. Most soy in North America is genetically modified. As with milk-based infant formula, parents know that their infants need to consume these products, regardless of the taste: "taste comes last." After all, medicine doesn't taste good. As the ad for a cough mixture admits, "It tastes horrible and it works." Parents assume that if infants cry, protest, gag, spit up or refuse to consume a soy-based formula, it is because they do not like the taste.

In parenting blogs, parents express concern at having to force the commercial substitutes into their infants, fearing the health consequences if all infant formulas are rejected. This fear is exacerbated by movements such as the fed is best foundation (fedisbest.org). The first page of the mission statement features the words of the medical doctor who co-founded the foundation: "I am absolutely pro-breastfeeding and absolutely AGAINST starving a child to achieve it." To suggest that breastfeeding advocates would rather see parents starve their children than provide them with these commercial alternatives is patently absurd, dangerous, and an insult both to parents who use these products and to breastfeeding mothers who need support.

Field Logic: Taste Talk Around the Table

Sensory science has provided suggestive evidence about the importance of infant feeding. Lab logic suggests that breastfed infants become more interested in new tastes and less likely to become picky eaters when they become older, compared with infants who were fed milk or soy-based products. Breastfed infants expect constant changes in the taste of food. But there are other routes to food acceptance, and other ways to explore how tastes develop. A conversation around the table provides some anecdotal evidence of other directions to explore. It is worth noting that before these conversations began, initial inquiries among adults in a southern Ontario town revealed little interest in and often no knowledge about how they were fed as infants, a further demonstration of the difficulties in studying this topic ethnographically. Consequently, I begin by reflecting on my own taste career.

I hate the taste of fish and seafood, although my mother loved fish. I can remember the first time I tasted a tuna fish sandwich and vomited. Ever since that experience as a four-year-old at a birthday party, I have avoided fish (but not birthday parties). I know I have a food aversion not a food allergy, but I cannot will my body to like fish. I have consumed it by accident; when it was pointed out to me that I ate fish, I immediately felt nauseous—but not before I had consumed a second helping of the raw tuna marinated in lime juice that I thought was chicken.

I cannot link this story to how I was fed as an infant, except that I was told that I would eat nothing but fish (caught by my father) the summer when I was two years old. I was never breastfed, but given a homemade formula from lactic acid milk and corn syrup, and have no conscious taste memories of that mixture. After chemotherapy in my sixties, I lost taste sensitivity but continue to crave intense sweet tastes and chewy textures; unfortunately, this includes Dutch fruit jellies, jujubes and gumdrops—a craving for sweet chewy textures I have had since childhood. When I was around five, I used to visit my bed-ridden uncle who kept boxes of square cut jellies rolled in sugar in his bedroom. My mother never figured out why I was so willing to visit that uncle and not healthier relatives.

My daughter was breastfed for sixteen months, but given infant formula by bottle every Thursday during school terms while I taught a double session of anthropology tutorials. We gave her rice gruel at night around two months of age hoping this would encourage her to sleep through the night. It didn't. When she began eating foods more regularly, she wanted foods she could pick up and feed herself. She did not like being spoon-fed. We used a small portable food grinder to prepare her food from our meals. Around her first birthday we travelled to Thailand for fieldwork, and were concerned about her meals. We took care to prepare her meals carefully using bottled water to make rice soup with chicken legs and fish balls purchased from foreign supermarkets. That is, until we saw her on the floor with her babysitter eating fermented fish paste and chilies with her fingers.

I always taste the food I give to others; my breastmilk was a notable exception. I had neither the inclination nor the opportunity to taste my milk, but always a lingering curiosity. Hence my interest in Jess Dobkin's performance art piece, the Lactation Station Breastmilk Bar, first performed in Toronto in 2006 (Van Esterik 2008), where participants had the opportunity to taste human milk donated from several mothers with different diets. Some visitors had a disgust reaction to the idea of tasting human milk; others an obsessive fascination with the idea.

People often have a disgust reaction to cheese and ice cream made from human milk not necessarily because of the taste (few have tasted human milk in any form), but the idea that the wrong person was eating the product; entrepreneurs were taking the food designed for an infant and giving it to an adult for no good reason other than making money from the shock value. After all, they were not using human milk medicinally to cure eye diseases or to nourish a person with AIDS in the days before therapies were available.

Others around the table recalled memories that shaped their food preferences. Marriage changed one relative's food preferences, training him to eat the squash he disliked because his wife liked it. A cousin recalled being told she was a good eater as a baby, adjusting easily to a milk-based infant formula. She advanced from gruel to commercial jars of baby food at a few months of age.

Another cousin recalled being told she was a "restless eater." She could not tolerate milk-based formula and was soon started on Sobee and Mull soy formula. She refused to eat cereal or peas, but loved fruit. I recall watching as her exasperated mother tried to force her into eating peas and gruel, to no avail; she just would not open her mouth. She identified her child-self as a picky eater, unwilling to try foods she did not like. She always had problems with dairy foods, but her baby records show that she was gradually forced to accept milk by starting out with mild cheeses, cottage cheese, and progressing to three glasses of skim milk a day. As an adult, she suspected she was lactose intolerant, but her tests never showed a true milk allergy; she just never enjoyed dairy products. After asthma and allergy problems, she cut out all dairy on her own and her allergic symptoms rapidly disappeared. Today, she enjoys peanut butter, never lost her sweet tooth, but developed a dangerous pineapple allergy resulting in anaphylactic shock later in life.

Her memories of feeding her sons are not pleasant. Her first-born son was induced and the drugs made her son sleepy and jaundiced, unable to suck. She received no help in the hospital but recalls a low point when a group of interns came in to her room to set up an IV for a glucose drip and to take a blood sample in the arm she was using to hold and encourage her reluctant sleepy newborn to latch on. A beautiful, full-breasted redhead, she cringes decades later to recall the interns ogling her exposed breasts. She had difficulty finding any formula that agreed with her son George. After several attempts with milk formulas, including SMA, and even boiled goat's milk, he was put on Nursoy. When that failed to agree with him, he was put on a hypo-allergenic formula (Nutramigen). To no avail; he spent the first year of life with constant diarrhea, and low weight gain, although he was receptive to new foods. Eventually he began to gain weight, and settled down with only a few food dislikes that persisted into adulthood—tomatoes, eggplants, olives, and milk. He loves cheese and his favorite comfort foods include macaroni and cheese, grilled cheese sandwiches, and shepherd's pie. Son George remembers none of this, except that he had a problem with milk when he was an infant. He doesn't link his adult taste preferences to any specific events, but to "something that put me off."

George's brother, Fred, had no digestive problems with the milk formula he was given, but he had constant skin rashes from birth, and was quite "shaky," comforted by constant carrying. Like his brother, he ate household foods, and was receptive to unusual tastes such as lentil soup at an early age. He never exhibited signs of food allergies or sensitivities, except through his sensitive skin. Unlike his older brother George, second son Fred loves to drink milk.

The table produced other stories of siblings who eat opportunistically. What one sibling prefers, the other rejects; what one sibling craves, the other finds disgusting—bananas, milk, fish. Sam's early experience with eating uncooked onions resulted in a long-lasting dislike of foods with hard textures. Speaking of food aversions, Joyce recalled her experience of the meal of cabbage she ate before she went into labor that she regurgitated throughout labor, creating an aversion to cabbage that lasted for years.

Our neighbor Taylor, a vegetarian for most of her life, is proud of the eating habits of her thirteen-month-old breastfed son, Allan. After her Caesarean section, she was unable to breastfeed right away and purchased a very effective double electric breast pump. Her dislike of cow's milk is reflected in her son's diet. He has never tasted cow's milk but enjoys organic almond milk along with a full diet and his mother's pumped breastmilk. She has exclusively pumped since he was around seven months old when he began biting and refused to be breastfed. Around five to six months, Allan was fed organic rice cereal mixed with breastmilk and took to solid foods rapidly. He prefers to feed himself with the same foods his mother eats—humus, guacamole, black beans, squash, blueberries, bananas, and kale—a wide range of vegetarian foods. He has never been given any refined sugar products, and rarely eats convenience foods. He enjoys spicy food and curries but has a disgust reaction to peas. His father eats meat, and his mother tried to retrain her palate to eat meat but did not enjoy it. No doubt Allan may also want to experiment with meat, but for now, he is a healthy vegetarian who tries new foods and likes to be in control of his food choices.

Taylor was not raised vegetarian but changed her diet for health reasons as an adult. As a child, she never enjoyed drinking milk. Vegetarian mothers often assume that their young children will not enjoy the taste of meat. Stories circulate among vegetarians that their babies will spit up if anyone tries to give them meat. It would be interesting to know if the child of a vegan mother who never tasted meat *in utero* is likely to enjoy the taste of meat.

Science and Stories: Broadening the Conversation

Infant constitutions are shaped by many food-related experiences. Infant feeding provides an opportunity to explore how taste preferences are formed interactively first with the mother, then with other caregivers. As the infant moves into larger commensal circles, the social construction of taste becomes more visible. Circles of food sharing expand to include siblings, other family members and care givers, and schoolmates. Of course, parents may not always know what their children consume (cf. Wilson 1974).

There are differences between tasting in labs and tasting in social situations. Lab logic tests for flavor compounds, while people eat mixtures of foods in social settings.

The stories around the table suggest a few additional routes to understanding the food and taste preferences of children in addition to how they were fed as infants. These include:

- over-exposure to a single food at a critical period of sensitivity to taste;
- food textures and odours;
- eating in competition with others;
- life-changing events like marriage, childbirth, and death;
- food etiquette and the desire not to offend others by rejecting food;
- memories of special food-focused events such as birthday parties, weddings, potluck suppers or picnics.

Tastes change over time. Events that change the composition of the household alter the household meal system. The stories suggest that changes in taste might best be observed at points of transition such as birth and marriages when taste perceptions are negotiated, leaving the eater with long-lasting taste tensions. Social experiences such as relations with siblings may have an impact on food preferences by encouraging eating in opposition to others. Sibling rivalry can encourage competitive eating in Euro-American households, while in other parts of the world such as Southeast Asia, sibling caregiving may encourage preferential food sharing, with the elder sibling encouraged to give the best bits to a younger sibling (cf. Van Esterik and O'Connor 2017).

Lab logic provides evidence about taste transfer, but food socialization is also shaped by culturally constituted foodways, including expectations about food items, etiquette and meal formats. Culture acts as filter for taste. Consider, for example, the contrasts between French and North American meals. Work by Rozin and colleagues (2011) demonstrated how the French preference for communal values, quality, moderation and joy contrasted with American preferences for personal values, quantity/abundance and comfort/convenience, expressed through food. American meals are about taking personal responsibility for nutrition and health; French meals are more about the sociality of meals and a shared value of what tastes good. As a result, French consumers expect less variety and choice compared to Americans.

North American children in food secure households may have little experience in eating food that they dislike the taste of; even adults may choose to only eat foods they know and like. Is it important to keep eating foods when one does not really enjoy their taste? Children may eat foods they dislike in imitation of a caregiver, or for reasons of etiquette. For example, French toddlers and young children in day care are served three course meals, and when they encounter foods they do not know or like, they are encouraged to just eat a little, and leave more on the serving dish for the next child who may like it better. This early training in commensality based on traditional cuisines may help balance out cravings and aversions (Shepherd 2012: 236).

Taste, individually developed through the maternal body but reflective of the eating experience (including food security) of earlier generations, gradually becomes collective through the accumulation of experience of specific culturally constructed flavor regimes presented in familiar meal formats. But the flow of tastes between maternal and infant bodies through breastfeeding is clearly only one piece of the story. While the taste perceptions of adults are influenced by language, cognitive preparation, nostalgic memories of smells, and what they expect or know about the food product they are about to taste; infants do not have that experience except from *in utero* memories. It may not be possible to distinguish between flavor images formed *in utero* and food memories acquired by other early eating experiences. Taste memories from the womb may not be retained when there is no maternal milk transferred, as in my case. Perhaps I consumed rotting fish as a toddler, and that shaped my aversion to all fishy tastes. I have no memory of the event and no one to ask. Infants and young children may lack memories of traumatic food events. Sam's experience of encountering raw onions in a dish turned him off the textures of hard food. He may not remember the incident but his mother does. This is a reminder that it may not be the taste of food but the texture or smell that is remembered. But the gut remembers. Recall the meal of cabbage that Joyce consumed before labor began and regurgitated through the long hours before giving birth.

Lab logic and field logic provide a few pieces of the puzzle to suggest how maternal breastfeeding and commercial infant formula might set the taste palate for future food preferences. Stories around the table suggested additional routes linking infant feeding and adult food preferences that could guide future focused ethnographic research on this subject. After all, my kitchen table hardly qualifies as a field site. But this chapter has not provided definitive evidence for the link between infant feeding and adult food preferences; the path is not direct or linear, but there are hints about the lasting impact of infant feeding regimes from both science and stories. Food transitions of infancy are mediated by maternal food preferences and household food practices, regardless of what infants are fed.

Some infants experience their first tastes from a familiar embodied source, human milk, a product that carries the micro-distinctions of embodied *terroir*; every mother's milk tastes different. Just as the taste of artisanal cheese depends on seasonality and what the cow ate, so too human milk carries traces of maternal meals. If taste enacts identity, what are the implications of consuming the embodied *terroir* of mother's milk vs the industrial taste of a commercial milk or soy-based infant formula? What does it mean for infants to be socialized into industrial tastes, or staple foods that "taste like poo"?

The Industrial Palate

> The new powdered instant baby food recently introduced by Heinz raises the possibility of another disturbing scenario. A baby brought up on a diet of formula made with powdered milk who then moves on to baby food made with powdered ingredients could, in theory, graduate to instant potatoes, instant soup, instant gravy and frozen TV dinners, and never taste fresh food.
>
> M. Burros, 1985, cited in Bentley 2014: 132

There is an underlying recognition in this speculation that the industrial palate is developed very early in life and acclimatizes a person to prefer tasteless processed industrial foods full of sugar, salt, additives, and saturated fats. Critiques of the industrial food system seldom extend their arguments back into infancy, with the important exception of Amy Bentley's book, *Inventing Baby Food* (2014).

North Americans may be experiencing the loss of taste as a corporeal way of knowing the world. Several generations of infants never tasted the variations of taste from mother's milk, but instead experienced the industrial taste of processed food through varieties of infant formula shortly after birth. These children may be unable to distinguish between artificial and real flavors in food unless they are explicitly taught. The fat, salt, sugar, and artificial flavors in processed food may overpower the complexity of the taste of real food.

Breastfed infants receive constantly changing tastes of the table, while most milk-based formulas taste the same, and, as I have argued, some taste bad. Do infants fed with these industrial products get bored with the same taste at every meal, day after day? Do toddlers who overeat like the taste of food too much? Shepherd suggests that bottle-fed infants may overeat after they are full in an effort to find pleasure from a missing taste and escape a bland, boring diet (Shepherd 2012: 189). Some research suggests that obese individuals may have a higher sensitivity to sweet and salty tastes (Hardikon et al. 2017; Feeney 2017). Consider too the impact of constantly changing between different kinds and brands of formula. Does the gut, or more precisely, do the gut microbes, remember this as they scramble to adapt to new products?

Although infants thrive on a single food, milk, Americans in particular value choice and dietary diversity. Most people in the rest of world probably consume fairly monotonous diets based on a consistent staple accompanied by dishes reflecting seasonal variety. It took some time for me to adjust to rice as a staple of most meals in Thai and Lao villages, having grown up with the artificial and superficial diversity of branded foods in North America. But to explore these questions, we must place infant feeding into the broader context of the industrial food system, and consider the evidence for the addictive properties of salt and sugar in industrial food products, including commercial baby foods.

Sensory science studies helped produce industrial food and consult to make industrial food taste better. Currently, research on taste transfer and infancy is used by the baby food industry to improve the taste of infant formula and baby foods, and to manufacture constituents like human milk proteins to "humanize" infant formula. Although researchers deny that financial ties to the food industry influence their research, potential conflicts of interest should be disclosed. A recent editorial in the Lancet (2017) summed this up: "…from tobacco, to sugar, to formula milk, the most vulnerable suffer when commercial interests collide with public health".

Wealth in many societies means having enough to eat. How does taste socialization relate to food security and income disparities? If you are hungry enough, does taste matter? If infants are hungry enough, will they devour milk that smells like "dead rotten salmon" to an adult nose? Or will they refuse to take a bottle of commercial infant formula? This is a common fear in Euro-American families where bottle feeding with infant formula is a common back-up plan for parents. This fear that breastfed babies will refuse a bottle with formula can motivate the early introduction of supplementary bottles. In food secure communities, infants may wait out adults until they "get what they want"—mothers' breast. The "fed is best" foundation plays off these fears among mothers who have problems with breastfeeding. Taste is generally missing from these discussions. Generational changes in taste preferences are usually attributed to marketing of new industrial foods, without considering the taste socialization of infancy that produces children who come to prefer the taste of new industrial foods. Poor infant feeding practices several generations back may also contribute to infant feeding problems in subsequent generations, taking some pressure off parents and mothers. This topic can be explored without adding an extra layer of mother-blaming to choices made in situations where the necessary support is not always there for families, especially breastfeeding mothers. By rephrasing the question around how taste becomes public, we can avoid the bitter aftertaste of morality.

References

Bentley, A. (2014). *Inventing Baby Food*. Berkeley: University of California Press.

Bourdieu, P. (1984). *Distinction: A Social Critique of the Judgement of Taste*. London: Routledge and Kegan Paul.

Burnier, D., L. Dubois, and M. Girard. (2011). "Exclusive Breastfeeding Duration and Later Intake of Vegetables in Preschool Children." *European Journal of Clinical Nutrition* 65: 196–202. daddytypes.com/2004/06/20/baby_formula_taste_test.php. retrieved February 24, 2017.

Debucquet, G. and V. Adt. (2015). "The Naturalist Discourse Surrounding Breastfeeding among French Mothers." In *Ethnographies of Breastfeeding*, eds., Tanya Cassidy and Abdullahi El Tom, 79–98. London: Bloomsbury.

Dettwyler, K. (1994). *Dancing Skeletons: Life and Death in West Africa*. Prospect Heights, IL: Waveland Press.

Freud, A. (1946). "The Psychoanalytic Study of Infantile Feeding Disturbances." *The Psychoanalytic Study of the Child*, vol. 2.

Hardicon, S., R. Hochenberge, A. Villringer and K. Ohla. (2017). "Higher Sensitivity to Sweet and Salty Tastes in Obese Compared to Lean Individuals." *Appetite* 111: 158–165.

Harris, G. and Coulthard, H. (2016). "Early Eating Behaviours and Food Acceptance Revisited: Breastfeeding and Introduction of Complementary Foods as Predictive of Food Acceptance." *Current Obesity Reports*, 5(1): 113–120.

Hausner, Helene, et al. (2009). Breastfeeding Facilitates Acceptance of a Novel Dietary Flavour Compound. *e-SPEN, the European e-Journal of Clinical Nutrition and Metabolism*, 4(5): e231–e238.

Hoegg, H. and J. Alba (2007). "Taste Perception: More than Meets the Tongue." *Journal of Consumer Research* 33: 490–498.

Klein, M. (1928). "Early Stages of the Oedipus Conflict." *International Journal of Psychoanalysis*, vol. 9: 169–180.

Korsmeyer, C. (1999). *Making Sense of Taste*. Ithaca: Cornell University Press.

Madrelle, J., C. Lange, I. Boutrolle, O. Valade, H. Weenen, S. Monnery-Patris, S. Issanchou, and S. Nicklaus. (2017). "Development of a New in-Home Testing Method to Assess Infant Food Liking." *Appetite* 113: 274–283.

Mennella, J. and S. Castor. (2012). "Sensitive Period in Flavor Learning: Effects of Exposure to Formula Flavors on Food Likes during Infancy." *Clinical Nutrition* 31(6): 1022–1025.

Mennella, J., et al. (2009). "Early Milk Feeding Influences Taste Acceptance and Liking During Infancy." *American Journal of Clinical Nutrition* 90: 780S–788S.

Mennella, J., C. Jagnow, and G. Beauchamp. (2001). "Prenatal and Postnatal Flavor Learning by Human Infants." *Pediatrics* 107(6): 1–6.

Mosby, I. (2009). "'That Wan Ton Soup Headache': Chinese Restaurant Syndrome, MSG, and the Making of American Food, 1968–1980." *Social History of Medicine* 22(1): 133–151.

Perrine, C. et al. (2014). Breastfeeding Duration is Associated with Child Diet at 6 Years. *Pediatrics* 134: s50–s55.

Rozin, P., A. Remick, and C. Fischler. (2011). "Broad Themes of Difference between French and Americans in Attitudes to Food and other Life Domains: Personal Versus Communal Values, Quantity Versus Quality, and Comforts Versus Joys." *Frontiers in Psychology* 2: 177.

Sausenthaler, S. et al. (2010). "Effect of Hydrolysed Formula Feeding on Taste Preference at 10 Years: Data from the German Infant Nutrition Intervention Program Plus Study." *Clinical Nutrition* 29(3): 304–306.

Sears, W. (1999). *The Family Nutrition Book: Everything you Need to Know about Feeding your Children—From Birth to Age Two*. New York: Little, Brown.

Seremetakis, C. Nadia. (2005). "The Breast of Aphrodite." In *The Taste Culture Reader*, ed., C. Korsmeyer, 297–303. Oxford: Berg.

Shepherd, G. (2012). *Neurogastronomy*. New York: Columbia University Press.

Wilson, C.S., (1974). "Child Following: A Technic for Learning Food and Nutrient Intakes." *Journal of Tropical Pediatrics*, 20(1): 9–14.

Wilson, B. (2015). *First Bite*. New York: Basic Books.

Van Esterik, P. (2008). "Vintage Breastmilk: Exploring the Discursive Limits of Feminine Fluids." *Canadian Theatre Review*, 137, Winter, 20–23.

Van Esterik, P. and R. O'Connor. (2017). *The Dance of Nurture: Negotiating Infant Feeding*. New York: Berghahn Books.

3

The Taste of Intervention

Tasting, Eating and Feeding after Weight Loss Surgery

Bodil Just Christensen and Line Hillersdal[1]

Introduction

This chapter is about the role of taste and tasting among a group of women who have undergone gastric bypass, an irreversible surgical treatment for severe obesity. It focuses on how they handle changes in their diet after surgery and the hard work many of them put into complying with the guidelines induced by surgery while still caring for their own needs and the needs of their families. We show how taste and sensorial engagements are integral parts of both the intervention and its effects. We interviewed more than fifty women going through gastric bypass surgery from 2011–2017 in Denmark. We interviewed them before and after surgery, accompanied them on their visits to the dieticians and hospitals and visited them in their homes.[2]

Exploring their struggle to start eating again after surgery and how they slowly adjust themselves to their changed bodies, we found that taste is not just the physical tasting of a food item but rather a range of food-related sensory and social practices that extend beyond an individual body and are shared within the family or social unit. Even before surgery, but notably after surgery, patients are advised and instructed to adopt new preferences and they start to work on developing a taste for specific food items. The health professionals frame achieving and maintaining a long-term weight loss as dependent on such changes, e.g. to manage hunger by choosing protein-rich foods and whole grains, or manage a "sweet tooth" with nuts or meal replacement bars, and to leave "old, bad habits" behind, which in practice means avoiding eating fatty and sugary foods. Through this learning process we suggest taste becomes disputed and challenged, and the bodily anchored, sensory engagement with foods and their tastes is transformed into new ways of relating to and prioritizing food and eating which have social implications.

Since the majority of patients are wives and mothers eating post-surgery affects not only the individual patient, but a range of significant others. The women we met are central figures in their households as family caregivers and providers of food. Often they and their families share lifelong experiences with obesity, which belong intimately to who they are as families (Grøn 2017). And though obesity surgery is often conceived of as an individual treatment for severe obesity, the intervention has many and often unexpected consequences for the households.

By exploring the social, material and practical consequences of surgery and the ensuing eating guidelines, this chapter argues that obesity surgery patients confront taste as a specific treatment methodology—the "taste of intervention."

Bourdieu (1984) was one of the first to argue that taste is not inherent to the object itself, but that taste is part of social distinction processes. We approach eating as an embodied practice, through which norms are produced and reproduced, and cultural classifications and social imperatives are adopted, thereby manifesting distinctions in taste and lifestyle (Bourdieu 1984; Levi-Strauss 1970). Learning new preferences challenges and makes visible this black box of naturalized norms and taken-for-granted practices and conceptions. On a more general level we comment on how food-eater interactions are expressed through constant "boundary making." When we eat we not only incorporate food but also a complex set of social relations (Falk 1994, Seremetakis 1992)—in other words, we make, remake and lend meaning to norms and social relations through eating.

The women—and thereby often their families too—have to deal with new preferences and tastes which (re-)position them in a broader socio-normative context permeated by strong narratives on health, weight and well-being. Having surgery for many patients also implies a deep sense of gratitude and appreciation, which in turn entails a strong commitment to respect dietary guidelines and health-related recommendations after surgery. For some this means that concerns related to health, e.g. fat percentage or fiber content, are deemed more important than tastiness. Hence, former taste regimes guided by pleasure are replaced by new ones guided by health. The ways in which tasting, or sensing the work that food does, between persons and dietary interventions, or between persons and their peers stresses the normativity of tasting and the ways eating has become politicized—what has been termed "a politic of the senses" (Farquhar 2002) connecting the bodily sensorium to modes of social ordering. Hence adopting new tastes and preferences comprises a larger social process on behalf of the women and their families where negotiations of shared identity and belonging are central to food buying, cooking and serving.

Taste as a phenomenon evolves from everyday practices, socialities and relations between food and eaters, and the lived experience of eating and tasting engages this interrelationship. To support this, we find helpful Hennion's account of taste as a social practice and a "modality of attachment to this world" (Hennion 2001, 2005). Through this theoretical lens we look at how the women deal with changing their tastes, preferences and diets and we ask how relations to significant others change as their eating changes.

The Gastric Bypass Procedure

Obesity rates have been dramatically increasing for the past fifteen years (Malik et al. 2013) and surgery has become a widespread and acceptable "treatment" for obesity.[3]

For the women we interviewed, surgery was an attempt to (re)gain control of their weight and eating. Contrary to other weight-loss procedures—which aim to change the body through lifestyle changes—surgery works by altering the body so that lifestyle changes are enforced. The most common operation performed is Roux-en-Y gastric bypass which consists of an elaborate procedure, where the stomach size is reduced and the length of the small intestine is shortened. The surgical procedure has both anatomical and hormonal effects: The reduced volume of the stomach induces fullness faster and a marked change in appetite hormones supports the sensation of satiety. In the obesity clinic the women are told that they can expect to lose approximately 60 percent of their overweight, but this varies greatly; some patients lose a lot of weight, while others lose less. In general the effect of surgery differs considerably between persons, and an intriguing feature in this context is that a small number of patients experience taste changes after surgery (Hillersdal et al. 2016). Neither clinical staff nor research on the subject (e.g. Benson-Davies and Quigley 2008) can fully explain the phenomenon and symptoms vary: for some patients all foods change and they reorganize their entire eating pattern or in severe cases lose their appetite, while others dislike just a few foods, e.g. meat, particular spices or chocolate. Another characteristic of the operation is the yet unexplained fact that in most cases patients are cured of their type–2 diabetes overnight (Rubino et al. 2016).

Another issue that concerns both medical staff and patients is that surgery does not provide a long-lasting solution to weight struggles—it merely provides a tool for a genuine lifestyle change. One of the doctors working in the ward mentioned that many of the patients he met had never had a normal way of eating, and had been either dieting or over-eating for most of their lives. This raises the interesting question as to whether a normal way of eating exists. In the ward the medical staff measured appetite quantitatively by using a so-called visual analogue scale (VAS) and some of the results were puzzling, e.g. when patients reported being equally full and hungry at the same time or never full (Christensen et al. 2017). Many of the women had lost a lot of weight at different points in their lives but they were not able to sustain the weight loss and consequently they invested a lot of hope in the procedure. But that said patients cannot expect to attain normal weight (Wood et al. 2016), and there is also a well-established risk of re-gaining weight after a few years (Magro et al. 2008). The effect of surgery, especially the hormonal changes increasing satiety, abates over time and surgery for some provides a tool and opportunity to change the lifestyle that led to the overweight (Hillersdal, Christensen and Holm 2016). This circumstance is a substantial element of the dietary learning program that patients undergo before surgery, as well as the central principles of the eating regimen after surgery.

There are four times more women than men receiving the operation which confirms the gendered bias in relation to both fatness and obesity surgery. In analyzing the pressure on women in the obesity discourse Throsby (2007, 2012) termed obesity a "woman's work." It is a recurrent theme in the literature that gendered norms of slimming and responsibilities in relation to the health of the family belong to women (Bordo 1993; Gough 2007; Broom and Warin 2011). The women seeking obesity surgery have all been overweight for large parts of their lives and were very concerned with their own overweight. Many of them see surgery as a critical tool for losing weight and avoiding future illness. But an important aspect is also the hope to "be normal" through attaining "a normal body size." It was remarkable how many of the women mentioned how losing weight could help them become a better parent, not just because they would feel more fit to play with their children but because they wanted to look normal to other people. Most of the women lived in nuclear families with children and others had adult children, while a few were single parents.

The family unit already plays a central role in the reproduction of societal norms including taste preferences (Bourdieu 1977; Ochs et al. 1996). It is well established within the literature on women, food and families that the shared meal constitutes the social unit as "family" (DeVault 1994; Murcott 1982, 1983). The pleasure of eating and serving food has been analyzed as a metaphor for the structure and emotional tone of family relationships (cf. Ochs et al. 1996) and is both a medium for providing care and a means for surveilling and exercising power (Wilk 2010).

The Taste of Intervention

After surgery women have to change their daily eating routines such as pace, meal pattern, food choice and sensing fullness. The women are advised and encouraged by dieticians and other health professionals in the bariatric clinic to schedule meals and control their eating in accordance with specific guidelines concerning nutritional content and eating patterns. These guidelines make up a comprehensive set of rules for "eating properly" post-operatively. A notable feature is that taste is not even mentioned in the written guidelines, mainly because the dietetic discourse stresses other characteristics of the food, such as calorie content, nutritional value, or digestibility, however, during counselling or patient education taste came up spontaneously in the dieticians' talk. The guidelines are well-established and generally unquestioned both among patients and care providers. Clinical staff use terms such as "lack of adherence" or "dietary compliance" (Harbottle 2011) to address the extent to which patients adopt the prescribed diet. Besides finding and adjusting to new routines with more meals a day the women are obliged to be very conscious of what and how much they eat, which is also new to some of them.

In patient education classes, before surgery, patients are shown PowerPoint slides with drawings of the surgical procedure and explained how it imposes a number of dietary obligations. The surgeon explained that the shortening of the small intestine decreases the uptake of nutrients, and that this decrease in combination with radically smaller portion sizes meant that foods preferably have to be as nutrition dense as possible in order to avoid nutritional deficiencies (Xanthakos 2009). The dietician proclaimed in a stern tone of voice that malnutrition is a very serious matter. She listed the recommended supplements and continued with the dietary guidelines after surgery. They dictate proteins at every meal, and a relatively high proportion of fish, fruit and vegetables, whereas carbohydrates, whether bread, potatoes, pasta or sugar, need to be restricted. Every patient received a folder. It had guide cards with colored tabs indexing the chapters: "Bariatric surgery and complications"; "Nutrition plan"; "Diet in practice"; "Fluids and alcohol"; "Discharge care" and "Physical activity." The chapter on "Diet in practice" provided patients with recipes and diet plans with suggestions for which foods to eat.

The suggestions closely follow Danish middle class ideals and preferences: white meats and fish, wholegrains, and seasonal fresh fruits and vegetables (Maguire 2016). In this way not only the diet but the very character of food is changed through the intervention. First, eating is discursively framed not as a source of sensory pleasure but as "nutrition" (cf. Scrinis 2013), a practice that primarily aims to achieve bodily health. The "nutritional requirements" are presented as "physiological needs" in a science-minded logic where taste is secondary, but the practical advice and recommended foods are permeated by social norms and cultural constructs. An illustrative example is that diet sodas and other products with artificial sweeteners or highly processed foods such as meal replacements (which many of the women still prefer) are nutritionally in line with the recommendations but conflict with the ideological values and whole-foods-and-fresh-meals ideals which are implicitly articulated by the health professionals. In one of their talks they introduced full dark chocolate as an efficient way to deal with a sweet tooth. Expressing how the intensity in taste of the chocolate was filling and satiating meant that specific sensory experiences of taste were taken-for-granted but in fact not shared by many of the patients present (see Hillersdal, et al. 2016). The "nutritional requirements" are thus translated into classed eating and taste practices that dictate food choice that are not always sensible (or edible) to the women. The intervention thus introduces "new tastes" both practically and symbolically. During counseling and teaching sessions both before and after surgery dieticians help patients to reflect on and change their eating habits and start to work with "healthier" strategies. One of the ways they react to their new diet is to work on developing a taste for specific food items, e.g. switch white toast for rye bread or crisp bread rich in fiber; or replace fat meats with lean cuts; or swap sugary soft drinks with water; or have fruit and vegetables as between-meal snacks instead of candy and chocolate bars.

Hence the so-called "appropriate taste" as conveyed by the clinical staff is also socially modulated, and the sensorial as therapeutic tool promotes specific relationships between eaters and recommendations, which encompass not only sensorial aspects, but also ethical, political and social concerns (cf. Sanabria 2015) which underline the ways in which our senses are intrinsic to socialization and thus evaluation. Being part of a dietary intervention for these women entails sensing and interpreting signals from their bodies in new ways. An example would be a counseling session with the dieticians where patients would tell about their eating. During one of these sessions one of the women mentioned how she would normally eat toasted bread with chocolate spread in the evening and how she now felt a subtle pain in her stomach when she did not eat this bread. In the session the sensation and urge was re-interpreted as thirst by the dietician. The woman herself expressed puzzlement, but was open to these new interpretations. Hence she accepted that the task now was a matter of learning to interpret signals from the body in new ways and to connect them with new taste preferences. These re-interpretations facilitated by the dietician—partly taking over the responsibility of knowing and interpreting their patients' bodily signals—allow the women new experiences when dealing with food choice and meal routines which are explicitly normative.

Not only do the women interpret their needs in new ways and find their bodies react differently to earlier routines but they also find themselves judging the behavior of family members differently which suggests that it is not only the individual woman who becomes the target of the intervention but both her and her family.

Taste as a Social Practice of Care and Relatedness

The majority of patients come from working class families. Women with little or no education are employed full-time or part-time in low-income jobs, or receive social welfare benefits. The fact that eating, food-related practices, and taste preferences are closely associated with class comes through in the interviews. Traditional Danish dishes with fatty meats and no vegetables, or sugary snacks, soft drinks, and cheap fast food are an inherent part of daily routines of care in many working class families in Denmark (Smith & Holm 2010). Family habits often revolve around eating, but after surgery caring for the family by offering tasty food and treats is suddenly an ambiguous practice. Food they used to share in the family is no longer agreeable and meals the women would normally serve their families often do not conform to the dietary recommendations.

New preferences also have to be dealt with in the families, which in the process leads the women to re-position themselves vis-à-vis other family members, as when they start to cook new dishes, insist on *not* watching television during dinner or radically minimize the amount of sweets and cookies kept in kitchen cupboards and find themselves preoccupied with the weight of partners or children.

Many women explain how they have "family rituals" around specific foods: crisps are indispensable for an evening in front of the television, candy Friday, Danish pastries on Sunday mornings, deep pan pizza Friday nights to celebrate the weekend. These habits are valued by all family members and are a way of doing family that everybody appreciates and caring for the taste preferences of children and partners ties an intimate social bond in the family, but such practices are incompatible with the bariatric surgery patients' new eating regimen. As a consequence shared eating and taste practices in the family change after surgery and the dietary counseling sessions. Navigating these changes might be challenging as eating is also about expectation and the roles one plays as a provider. The mothers are thus confronted by a range of negotiations marked by both cherished habits and care, and compromises and conflict.

It is not just preferences that mothers need to take into account when preparing family meals. There is another dimension of sociality too. A mother of four now grown-up children explains: "I enjoyed doing all these things; my kids could bring friends home and I would cook all sorts of nice things, and so they would think it was really great to visit us. . ." But the social role of being the caregiver through food often changes after surgery, as this mother of two explains:

> I have always been praised for being a good cook and a good hostess, but now, given that I can't cook—oh, well, of course I can but I can't eat it and then the question is if I bother. . . We'll see. . . and how it [the surgical procedure] affects me. . . It is a part of my identity that I will need to face and figure out.

Others approach it slightly different:

> There are expectations when it comes to Christmas. They are used to me baking all sorts of cookies, and that I make many, many of them. And now it's limited, because I can't be bothered. I bake all of them but I can't eat them myself. So that's why I don't bake as many as I used to. But I do bake for them.

This woman retains her position and role as a caring mother making the expected selection of Christmas cookies, but she scales down the number of each variety and thereby the time she spends on baking, because it is no longer as relevant and important for herself. In this way she accommodates both her own and her family's needs.

Most of the women discovered that they began to prioritize differently. Earlier cherished values of being a good hostess and a good cook were no longer as important as before, or conflicting needs or interests were solved by other means, e.g. by cooking different dishes for each family member. Though most women still cook dinner for their families, it is not necessarily a central activity anymore, hence some start to hand over the cooking to other family members and some focus on new activities that tie the family together, such as roller skating or family trips.

Furthermore, their attention and awareness of health and weight made them more sensitive of their responsibilities as mothers and the broader socio-normative context of eating and feeding others. One mother said:

> Our youngest is only 15 and quite chubby. He is fond of food. Extremely fond of food and really lazy and doesn't bother to exercise. So I try—out of care for him—not to fill [the fridge] more than necessary to avoid tempting him. He wants to decide for himself what to eat so the selection of food should not be too great.

She deems it her responsibility to limit his choices and thereby try to stop his weight gain. The intervention in the clinic becomes an active part in the renegotiation of family routines which have implications for how family members are judged, but can also be empowering for the women as they suddenly find that they have a route laid out for them or motivation to act as the intervention makes them more aware of new strategies. This can, however, be overwhelming. One woman felt this particularly intensely. She had suffered heavily from nausea and vomiting for a couple of weeks. She had surgery six weeks prior to the sudden discomfort and discussed the issue with the doctors from the ward. She explained:

> I have found out that it's psychological that I have felt sick. I simply cannot sit and watch other people eat larger portions than me, and shovel it in. It makes me physically ill and I get a strong feeling of nausea. I throw up if my husband eats too fast or eats too big a portion. I've told him: "You have to take your time, just like I do when eating, otherwise it'll go wrong." I talked with somebody who had surgery a week and a half after me and she is also disgusted and also gets nausea. So it must be our minds playing tricks on us.

This woman's visceral reaction created a distance from her husband whose eating disgusted her. She later explained how she did not bother to cook his favorite dishes anymore. She eventually left him. This radical example illustrates the effect of these changes in diet and how what these women used to hold sacred is suddenly open to scrutiny.

Negotiating and Embedding New Taste, Diets and Practices

How mothers face dilemmas of conflicting priorities, very often over health, weight or comfort/nurture, are recurrent themes, both in our ethnography and in the literature. They manage these quandaries in different ways: Some change their own habits to accommodate others' preferences (both husbands and children) (Iversen and Holm 1999), others cook two or more meals to meet all desires or requests (Furst et al. 1996), while yet others try to change the preferences of their loved ones so that they become more in line with the post-surgery eating regimen.

The fact that many of the women have been dieting most of their lives means that they have been eating differently from the rest of the family at different time periods in the past. Then why does this situation post-surgery seem different? Internalizing the individualized approach, most of the women stress how they do not want their personal intervention and special diet to affect their families. Their new way of eating becomes more visible in the families and some also find that they just have to eat differently, so their new diet does not have the same character of temporariness as former diets might have had.

After surgery one woman pondered her husband's food choice:

> When we go to a restaurant he gladly eats his salad, but not when I make it, which I do because I should eat it myself. I actually do prepare something separate for them because they don't have to follow me [my diet]. I don't want to force them. So they can easily get a pork chop and have the salad on the side. But he immediately says: I don't want to eat the salad.

Similar considerations and critical scrutiny of taken-for-granted habits are recurrent, e.g. when asked which habits she wanted to change, one young mother said:

> For instance, that having a cozy time doesn't necessarily only include unhealthy stuff. In my childhood we never lived healthy so it's sort of a way of starting over in my family with my husband and my child. For example that there must be cream in the sauce . . . but there actually doesn't have to be. Or, that buns have to be all white and soft and full of butter to be birthday buns in the right way. But it may well be different, and that has actually been kind of an eye-opener. My son is in really healthy daycare with a zero tolerance for sugar. And I can see how all those kids just love it when there is fruit or beetroot muffins or pasta Bolognese with carrots and stuff when it is someone's birthday. If the kids can have a great time just eating this, then all people can. It is nonsense that we think we can only comfort ourselves with candy.

She continued by telling how she gradually changed family meals and tried to influence the tastes of her husband and son by adding more vegetables, leaving cream out of the sauce, and replacing regular pasta with whole-grain pasta. This is a common strategy and often supplemented with cooking more dishes or sides for the shared meal. Others stopped eating meals with the family—they still cooked for them but had their own meal before or after the rest of the family. Through these various practices, how the women chose to manage eating in their family after surgery thus reproduced, negotiated, changed, destabilized, weakened or even broke up the intimacies and social ties in the family. Thus surgery—though perceived as an individual treatment for obesity—affected both patients and families.

Taste as a Treatment Modality

Meaning and identity formation also come into play for obesity surgery patients. We claim that patients are socialized by health-care staff and dieticians to conceive of their food, taste preferences and selves in different ways post-surgery. Though obesity discourse and its moral repercussions have been heavily theorized (e.g. Bordo 1993; Throsby 2007), there are few studies exploring the mundane practices of "eating right" (Mol 2009). Being healthy has become a moral imperative (Lupton 1995) and when it comes to food choice what used to be considered private has become a public matter, as "good citizens" are held responsible for their own health and well-being, for instance through eating practices (Wright and Harwood 2012). Discourses and norms around healthy eating are distinctly classed—notably in popular cultural and media narratives depicting simple dichotomies between compulsive, gluttonous fat bodies and reflexive, thin ethical eaters (Guthman 2003). Hence the moral-health-class nexus determines that to be a decent person, it is necessary to be a healthy person, and a healthy person is a slim person. Such norms surrounding foodstuffs, taste practices, table manners and ordering of dishes can be and are managed in different ways by caregivers and their families. The most common outcome is that not only the surgery patient, but the entire family complies with the bariatric dietary guidelines. When interviewed a year or more after surgery many women explain how husbands and kids have followed suit:

> They follow me, everybody in our house follows the same principles. They help and support me. They want it too. In the beginning I cooked a small portion of something for myself, and asked them what they wanted. But they want to eat the same things as I do: meat, chicken, fish, and a lot of vegetables.

As mentioned earlier several patients experience taste changes (Hillersdal et al. 2017). Food simply tastes differently. But in our analysis there is more to taste changes. Patients often change preferences. Some start to buy organic or local produce. One woman mentioned that she has started to appreciate the freshness and smell of vegetables and fruit in a new way and focus more on the quality of what she buys than the quantity. In general they dissociate themselves from former eating habits and denounce what they term the "crappy food" that they used to eat. Saying things like: "Before I just ate, I didn't care" or "I don't consider junk food to be 'real food.'" We will argue that changes in these seemingly natural dispositions and sensory experiences are socially informed. Thus the experience of a particular taste, in everyday sensory aspects of eating, is linked to larger socio-political ethos of being a responsible citizen. The moral concerns of patients add an ethical dimension to the situatedness and complexity of food practices and taste preferences that have been discussed by food researchers in recent years (see e.g. Sutton 2010). The taste of certain foods—both "healthy" and "forbidden foods"—and changes in preferences are thus part of negotiations that comprise appetite hormones and hedonic responses as well as identity work that binds taste experiences to changing subjectivities (cf. Faquhar 2002). In this way taste and preferences operate as a tool for making moral concerns visible and translate them into tangible food practices, as when dieticians recommend wholegrain crisp bread and lean cold cuts, patients not only choose these foods they also favor them (Nielsen et al. 2017). What is deemed healthy and "appropriate" is thus also being considered desirable. Taste preferences are thus both the result and the reproduction of social norms and the display of moral competence among patients. In this manner the very taste of certain food items—commonly low-fat yoghurt, leafy greens, tuna, chicken, wholegrain crisp bread, almonds and nuts, etc.—enables patients to configure themselves as compliant and "good" patients/citizens.

Conclusion

The chapter has focused on the relations and practices which eating and taste create in families, and link to classed normativities and moral obligations. Eating has always been used to construct selves and bodies—and responsibility for one's health has increasingly come to lie with the individual, and the way bodies are equated with social values and moral characteristics (see Rose and Novas 2000). The normative pressure on individual women spills over into the families and their taste practices. Furthermore, their novel attention and awareness of health, weight, and well-being makes them more sensitive to their responsibilities as mothers and to the strong moral narratives surrounding eating.

They find themselves judging the taste preferences and bodies of other family members negatively, which disrupts earlier consensual agreements within families of preferences, pickiness and so forth. We find the intimate connection between eaters in a family central in understanding the "taste work" these women do.

The problem of obesity is often reduced to a question of maintaining the right energy balance and learning to control the body with willpower and does not address other concerns or motivations that make people eat what they eat. Other studies have stressed how interventions which focus on pleasure instead of control might help people lose weight and loosen the normative pressure (e.g. Vogel and Mol 2014). Many of these studies, however, still focus primarily on the person in the dietary intervention and not on the role of food and eating in a person's peer relations. We have focused on the way in which the intervention influences family routines and how women deal with issues of eating and feeding after surgery when they still have to take responsibility for feeding their families. This puts the focus on the embeddedness of the eater and the obvious difficulties in directing diet interventions towards the individual person. Furthermore, the analysis highlights how taste work is two-fold: the post-surgery dietary guidelines dictate specific taste preferences, which are heavily classed. One could argue that first patients internalize weight stigma, then they are compelled to internalize class sensibilities of taste. The taste of intervention thus targets much more than excess weight and presents an example of what Sutton has termed the "scientification of gustation" (Sutton 2006) where taste and eating are believed to be key to the social problems of the day and where taste is defined by narrow scientific principles rather than by broad cultural and family habits. The obesity surgery patients thus provide an illustrative case on how taste and taste practices, shared or individualized, are always embedded in everyday lives where both normative and political as well as gustatory and social dimensions have to be navigated.

Notes

1 The authors contributed equally to the manuscript.

2 The first period of fieldwork (2011–2013) was carried out collaboratively by both authors and the second period of fieldwork (2015–2017) was conducted by Bodil Just Christensen. The first fieldwork was part of a large interdisciplinary initiative, the *UNIK—Food, Fitness and Pharma* research (http://foodfitnesspharma.ku.dk/). Findings were then further developed during the second fieldwork as a part of Governing Obesity (http://go.ku.dk/).

3 Alongside the rising numbers of "larger than average bodies" there has been a growth in the number of surgeries worldwide (Buchwald et al. 2004). The increase has been dramatic in the Nordic countries, e.g. in Sweden the number of surgeries multiplied tenfold from 2005–2010 (SOReg 2015). Calculated as a ratio of surgeries done per number of citizens Scandinavia greatly outnumbers other European countries. This might be partly explained by the fact that healthcare is publicly financed and provided irrespective of income or private insurances in Scandinavia.

References

Benson-Davies, S., and D.R. Quigley (2008). "Food Aversions and Taste Changes Following Roux-en-Y Gastric Bypass Surgery." *Topics in Clinical Nutrition*, 23(4): 357–363.

Bordo, S. (1993). *Unbearable Weight: Feminism, Western Culture and the Body*. Berkeley: University of California Press.

Broom, D.H. and M. Warin (2011). "Gendered and Class Relations of Obesity: Confusing Findings, Deficient Explanations." *Australian Feminist Studies*, 26: 453–467.

Bourdieu, P. (1984). *Distinction: A Social Critique of the Judgement of Taste*. Cambridge: Harvard University Press.

Buchwald, H., Y. Avidor, E. Braunwald, M.D. Jensen, W. Pories, K. Fahrbach, and K. Schoelles (2004). "Bariatric Surgery: A Systematic Review and Meta-analysis." *Jama*, 292(14): 1724–1737.

Christensen, B., L. Hillersdal, and L. Holm (2017). "Working with a Fractional Object: Enactment of Appetite in Interdisciplinary Work in Anthropology and Biomedicine." *Anthropology and Medicine*, 24(2): 221–235.
DeVault, M.L. (1994). *Feeding the Family: The Social Organization of Caring as Gendered Work*. Chicago: University of Chicago Press.
Falk, P. (1994). *The Consuming Body*. London: Sage Publication.
Farquhar, J. (2002). *Appetites: Food and Sex in Post-Socialist* China. Durham: Duke University Press.
Furst, T., M. Connors, C.A. Bisogni, J. Sobal, and L.W. Falk (1996). "Food Choice: A Conceptual Model of the Process." *Appetite*, 26(3): 247–266.
Gough, B. (2007). "Real Men Don't Diet: An Analysis of Contemporary Newspaper Representations of Men, Food and Health." *Social Science & Medicine* 64(2): 326–337.
Grøn, L. (2017). "The Weight of the Family: Communicability as Alien Affection in Danish Family Histories and Experiences of Obesity." *Ethos*, 45: 182–198.
Guthman, J. (2003). "Fast Food/Organic Food: Reflexive Tastes and the Making of 'Yuppie Chow.'" *Social & Cultural Geography*, 4(1): 45–58.
Harbottle, L. (2011). "Audit of Nutritional and Dietary Outcomes of Bariatric Surgery Patients." *Obesity Reviews*, 12(3): 198–204.
Hennion, A. (2001). "Music Lovers: Taste as Performance." *Theory, Culture & Society* 18 (5): 1–22.
Hennion, A. (2005). "The Pragmatics of Taste," in M.D. Jacobs and N.W. Hanrahan, eds., *The Blackwell Companion to the Sociology of Culture*. Malden, Mass.: Blackwell Publishing.
Hillersdal, L., B. Christensen, and L. Holm (2016). "Patients' Strategies for Eating after Gastric Bypass Surgery: A Qualitative Study," *European Journal of Clinical Nutrition*, 70: 523–527.
Hillersdal, L., B. Christensen, and L. Holm (2017). "Changing Tastes: Learning Hunger and Fullness after Weight-loss Surgery." *Sociology of Health and Illness*, 39(3): 474–487.
Iversen, T., and L. Holm (1999). "Måltider som familieskabelse og frisættelse." *Tidsskriftet Antropologi*, 39: 53–64.
Lupton, D. (1995). *The Imperative of Health: Public Health and the Regulated Body*. London: Sage.
Maguire, J.S. (2016). "Introduction: Looking at Food Practices and Taste across the Class Divide." *Food, Culture & Society* 19(1): 11–18.
Malik, V.S., W.C. Willett, and F.B. Hu (2013). "Global Obesity: Trends, Risk Factors and Policy Implications." *Nature Reviews Endocrinology*, 9(1). 13–27.
Mol, A. (2009). "Good Taste: The Embodied Normativity of the Consumer-Citizen." *Journal of Cultural Economy*, 2(3): 269–283.
Murcott, A. (1983). "'It's a Pleasure to Cook for Him': Food, Mealtimes and Gender in Some South Wales Households," in E. Gamarnikow, D. Morgan, J. Purvis, and D. Taylorson, eds., *The Public and the Private*. London: Heinemann.
Ochs, E., C. Pontecorvo, and A. Fasulo (1996). "Socializing Taste." *Ethnos*, 61(1–2): 7–46.
Ochs, E., and M. Shohet (2006). "The Cultural Structuring of Mealtime Socialization." *New Directions for Child and Adolescent Development*, 2006 (111): 35–49.
Rose, N., and C. Novas (2000). "Genetic Risk and the Birth of the Somatic Individual." *Economy and Society*, 29(4): 485–513.
Rubino, F., D.M. Nathan, R.H. Eckel, P.R. Schauer, K. Alberti, P. Zimmet and S.A. Amiel (2016). "Metabolic Surgery in the Treatment Algorithm for Type 2 Diabetes: A Joint Statement by International Diabetes Organizations." *Surgery for Obesity and Related Diseases*, 12(6): 1144–1162.

Sanabria, E. (2015). "Sensorial Pedagogies, Hungry Fat Cells and the Limits of Nutritional Health Education." *BioSocieties*, 10(2): 125–142.
Smith, L.H., and L. Holm (2010). "Social Class and Body Management. A Qualitative Exploration of Differences in Perceptions and Practices Related to Health and Personal Body Weight." *Appetite*, 55(2): 311–318.
SOReg 2015: https://docs.google.com/viewer?url=http://www.ucr.uu.se/soreg/component/edocman/2016-06-09-arsrapport-2015-del-1/download
Sutton, D. (2006). "Cooking Skill, the Senses, and Memory: The Fate of Practical Knowledge," in E. Edwards, C. Gosden, and R. Phillips, eds., *Sensible Objects: Colonialism, Museums and Material Culture*, 87–118. Oxford: Berg.
Throsby, K. (2007). "'How Could you Let Yourself Get Like That?' Stories of the Origins of Obesity in Accounts of Weight Loss Surgery." *Social Science & Medicine*, 65(8): 1561–1571.
Vogel, E., and A. Mol (2014). "Enjoy your Food: On Losing Weight and Taking Pleasure." *Sociology of Health & Illness*, 36(2): 305–317.
Wilk, R. (2010). "Power at the Table: Food Fights and Happy Meals." *Cultural Studies—Critical Methodologies*, 10(6): 428–436.
Wood, G.C., P.N. Benotti, C.J. Lee, T. Mirshahi and M. Lent (2016). Evaluation of the Association between Preoperative Clinical Factors and Long-term Weight Loss After Roux-en-Y Gastric Bypass. *JAMA surgery*, 151(11): 1056–1162.
Wright, J., and V. Harwood, eds. (2012). *Biopolitics and the "Obesity Epidemic": Governing Bodies* (vol. 3). New York: Routledge.
Xanthakos, S.A. (2009). "Nutritional Deficiencies in Obesity and after Bariatric Surgery." *Pediatric Clinics of North America*, 56(5): 1105–1121.

4

The Taste of Reef

Changing Food Preferences and Taste in a Solomon Island Archipelago over Forty Years

Peter I. Crawford

This chapter explores the changes in food preferences and taste in a specific Pacific context, reflecting David Sutton's point (2010: 219) that while studies of changes in food habits are quite common, ethnographic cases in which the sensory aspects of taste are central are harder to find. I argue that in the case of the Reef Islands in the Solomon Islands, the social construction of taste has undergone significant developments over the past few decades, some undoubtedly due to taste being inextricably linked to food habits, which have altered mainly because of outside influences, but also due to the ways in which taste increasingly is something people, once a vocabulary is established for it, talk about and share in public.

In June 2015, I conducted a one-month follow-up fieldwork in the Reef Islands, Temotu Province, The Solomon Islands, the location of a long-term ethnographic film and research project that began in 1994. The 2015 fieldwork had a focus on food and nutrition, and was conducted with my wife, Birgitte Hansen, a Danish schoolteacher and nutritionist, with whom I managed a DANIDA[1]-funded development project, initially focused on nutrition, in the Solomon Islands in the 1990s. The main purpose of our fieldwork was twofold. First, to look into and document, using video, existing food consumption patterns in three different villages on three different islands (Ngasinue/Fenualoa, Nifiloli, and Pileni), comparing it with data from the fieldwork of fellow anthropologist, Jens Pinholt, and his wife, Kirsten Pinholt, a nurse, in the early 1970s, and a series of field and film work conducted from 1994 to present. Secondly, we were specifically interested in nutritional aspects related to the obvious and widespread hypothesis when it comes to the South Pacific, namely that it is a region thought to have moved from being one of the nutritionally most healthy areas on earth to communities increasingly suffering from such lifestyle diseases as obesity and diabetes, allegedly due to a diet brought about by becoming part of the globalized world. It is the former, with a particular emphasis on the notion of "taste," that is the focus of this chapter.

When I first went to the Reef Islands, Pinholt had warned me that one did not go to the Reef Islands for the culinary experience. I ended up completely agreeing, rarely having been anywhere in the world where to me the taste of food was as bland. The staple dried breadfruit, locally and nationally famously known as *nambo*, constituted the main proportion of most meals, usually supplemented by boiled root crops, such as sweet potato, cassava, or taro, and a small amount of fish sometimes at least adding some flavor to the meal. A more recent addition to any meal—although having been available even before independence in 1978—is imported white rice out of the ubiquitous yellow *Solrais* bags, which are frequently recycled as carrier bags after use. On festive occasions meat would also be served, chicken (*kokorako*) or, more importantly, because it is undoubtedly the festive food item par excellence, pork (*pikpik*). To the extent meals included vegetables it would consist of one of the many varieties of *kabis* (cabbage), either wild or grown, shallots and tomatoes. There was no evident use of spices or herbs, not even black pepper, salt often seeming to be the only condiment added. The only spice I noticed was turmeric, which, however, was exclusively used for ceremonial purposes as a coloring agent, of hair and body. Coconut milk was used, for example in so-called *pudim* ("pudding"), adding a highly welcome but very rich and fatty flavor to tubers. This was at a time when the Reef Islands still had remarkably limited contact with the rest of the world. What was more striking and interesting than my personal "taste" was the lack of expressions or discussions of taste, either in public or in the households we stayed in. Taste was thus not an issue, not something people discussed. It seems that the arrival of foreign food items, flavors, and exposure to people who did discuss "taste," such as us, helped trigger a discourse on taste. In that sense the question this chapter explores may almost be defined as: what are the origins of taste?

The Origins of Taste

Whether we can speak of origins of taste triggers two main critical observations. First of all, almost self-evidently, what do we mean by taste?[2] The, by now, vast body of literature on taste[3] indicates that a phenomenon with which we are all familiar, and have been so from birth, is not a straightforward matter. One of the early writers on taste, Brillat-Savarin (2017 [1825]), however, does emphasize a matter, on which all other writers seem to agree, namely that taste is multi-sensorial, i.e. involving all our senses, although he does separate taste as a sense in its own right, evident in his short definition: "Taste is the sense which puts us in touch with sapid bodies, by means of the sensations which they excite in the organ designed to appreciate them" (2017: 13).

While Brillat-Savarin, like many other early writers, does seem to dwell on physiological and chemical aspects of taste, albeit in a multi-sensorial context, a second critical observation in the context of Reef Island taste may be that we now realize how taste is also affected, perhaps equally so, by what Jack Goody (1982) has described as "culinary culture." This may historically be the most significant contribution of what increasingly has become known as an anthropology of food, namely that irrespective of the physiological and chemical similarities of taste, our culinary cultures help define what in a given context constitutes taste, and flavor(s). The implications of this are both spatial and temporal. At an individual level this helps explain why some people like this or that and others do not but also why we may, for example, hate mushrooms as a child (which I did!) but love them in our adult life. At a more general cultural level, which of course is what Goody was talking about, it explains why different cultures eat differently but also how culinary traits change over time, explaining why there is an obvious overlap between anthropological and historical studies of food culture.

Given the two critical observations above, it is obvious that my initial remarks concerning the lack of taste and blandness of Reef Island food say more about me and my food preferences, as is often the case, ethnocentricity slipping into the matter at hand. When the Pinholts first stayed in the Reef Islands in the early 1970s, and when I first arrived in the mid-1990s, most meals were probably bland to our northern European palate, but what did Reef Islanders themselves have to say about this and what have the major developments been in Reef culinary culture since then? "Taste" was, of course, always there but what changes have taken place and what may have triggered these changes? Answers to these questions may contribute to filling a gap in the literature on the Reef Islands that may have emerged simply because anthropologists and others writing about the islands did not find food interesting, or only to the extent that it formed part of trade patterns, food taboos in the context of matrilineal descent groups, or the role food has in the production of items needed to survive rather than ingredients of a culinary culture in its own right. Another reason may well be, as one of my older informants explained, that people rarely discussed food, certainly not in public, and expected anthropologists and other visitors to be interested in other matters. Certainly, the most important works, such as Davenport (1969) and Koch (1971), and almost embarrassingly our own work (Crawford and Pinholt 2010), seem by and large to neglect this whole field, the only anthropological article I have found dealing with food—but not taste—in the Reef Islands being Yen's (1975) contribution, based on William Davenport's work, to Margaret Arnott's seminal book (1975) on the anthropology of food. Even the kind of reports ubiquitously found in so-called developing countries, produced by development agencies and NGOs, are conspicuously lacking in the case of the Solomon Islands in general and Temotu Province in particular, a rather limited study from World Vision the only case I have been able to locate from recent years (Cowling 2008).

The Ethnographic Context of Reef Food

It often makes sense, and in the context of food culture even more so, to describe the two neighboring Reef Island communities as people of the land and people of the sea respectively. Fenualoa, reaching fourteen metres at its highest point, has sufficient topsoil to sustain intensive horticulture, as well as a "bush" being the home of a number of important wild food plants, such as breadfruit, bananas, mangoes, wild taro, and a number of nut species, apart from coconut palms, and areca palms that produce the cherished betel nuts, being the easternmost part of the betel nut consuming part of the world that starts in Southeast Asia. The five islands of Vaiakau, however, are small coral atolls, the sands of which may accommodate that iconic Pacific plant, the coconut

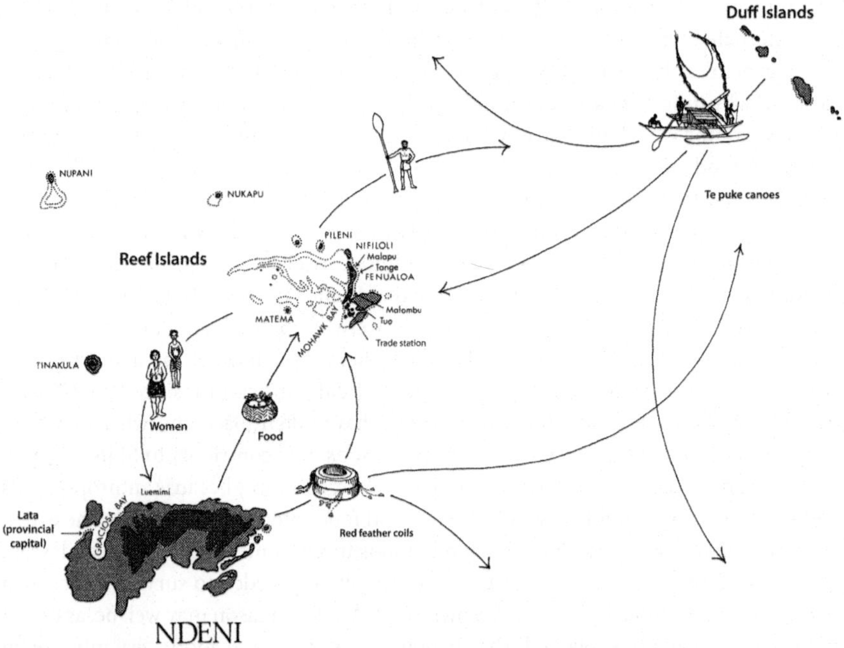

Figure 4.1 Drawing of traditional exchange system. Reproduced with kind permission from Tegnestuen, Moesgaard Museum.

palm, as well as bananas, some breadfruit and nuts, but apart from coconuts hardly even that on the smaller islands.

The densely populated Reef Islands as a whole have always relied on import of food items, mainly root crops, which have formed a central element in the traditional exchange system (see Figure 4.1). Within the Reef Islands, and the two neighboring ethnic groups, there seem to be no major differences or variations in food patterns, preferences and taste. Root crops and breadfruit are staple foods, dried breadfruit, *nambo*, having obtained an iconic status in the constitution of Reef Island identity. Paraphrasing Rozin and Rozin concerning what "makes Mexican food 'Mexican'" (2017: 37), chilli pepper, corn and beans, there seems no doubt that *nambo* and tubers are what make Reef food Reef, to the extent that they almost define what food is.[4] I have never heard a Reef islander express dislike for *nambo* and always returned from the islands to the capital heavily loaded with baskets of it, either as a gift to me and my family or for relatives in the capital, giving them the coveted "taste of home" (cf. Sutton 2017). What makes the comparison pertinent in the current context is that "Traditional flavors may serve the same function as traditional costume or traditional religious practice. They are a means of defining a culture group, of identifying an individual with it, and of separating that group from others" (Rozin and Rozin 2017: 39). In the Mexican case I believe most people, even non-Mexican, would agree that chilli does add a distinct flavor to the cuisine, whether one cherishes that particular flavor or not. In the

Figure 4.2 Dried breadfruit (*nambo*) for sale at Honiara Central Market, catering mainly for the relatively large Reef population that lives and works in the national capital (© Peter I. Crawford).

Reef case, however, even Reef islanders tell you that *nambo*, the most significant culinary identity marker, is almost devoid of flavor. One of my friends, Willie (see below), who had lived in other parts of the country, learning about "other" kinds of food, once said that although *nambo* had very little flavor, it was still what gave him the "taste of home".

One could possibly argue that what we are dealing with here refers to the distinction between taste and food preferences (cf. Macbeth and Lowry 1997), accepting that taste and flavor form only one dimension among others when it comes to what kind of food people prefer. However, one of my main points in the Reef case is that while possibly the most significant constituents of Reef culinary culture may be "tasteless" this may imply two things. First, that we must accept that, at least to some people, "tastelessness" in itself may define a taste (and certainly a food preference) and, second that taste is both a dynamic concept and cultural phenomenon that changes over time, even though this may not necessarily affect what is defined as a specific food culture. In a sense what I am arguing has to do with what may be described as the taste of place, but in a different, albeit linked to, sense than that based on *terroir* employed by, for example, Amy Trubek (2008). I am rather convinced that in the Reef case food "evokes 'narrating' place," as Norbye (2010: 147) describes it in her analysis of mountain summer dairy farming in Hemsedal in Norway. What defines Reef culinary culture is embedded in the tasteless *nambo* despite developments over time (and space) of Reef taste and the taste of Reef food.

The Taste of Reef

While the most emblematic food item representing Reef culinary culture, *nambo*, is arguably almost devoid of flavor, the culinary culture of the two neighboring communities has undergone substantial changes since the early 1970s, bringing in new tastes and flavors through increased contact with the outside world. While the following analysis is focused on material from three separate islands, Fenualoa, Nifiloli and Pileni, there is reason to believe that we can identify some general trends.

One of the difficulties associated with apperception of taste, leaving the matter of subjectivity aside, is the rendering of it in words. It is very difficult to reduce such a multi-sensorious practice to the realm of the logocentric. Even in languages such as English, which like many other languages, as Tuan notes, has an "exceptionally rich vocabulary of taste and texture" (2017: 226), verbal accounts cannot but delimit the exchange of taste sensation. We have always communicated with our Reef friends and informants in Solomon Pijin, the lingua franca of the Solomon Islands, only picking up a limited vocabulary from the two local languages Pileni and Aiwoo.[5] This may undoubtedly have affected our ability to gauge local definitions and descriptions of taste and flavor but on the other hand we have always had key informants capable of translating to English, according to whom the vocabulary of both local languages is limited when it comes to taste. In Pijin I have basically only come across five words specifically used in discussions of food appreciation and taste. The Pijin term *suit* is used in a much wider sense than the English word sweet from which it is derived. Although it may also denote sweetness it is more widely used to express that something simply tastes nice, the closest English equivalent probably being "delicious," and always used in a positive sense. To express degrees of whether something is delicious you use words such as *barava* or *tumas*, both meaning very. The only case in which I have heard *suit* used in a negative sense was in combination with the word *ova*, indicating that something is too sweet (not too delicious). Probably the second most commonly used term is *gris*, which, derived from the English word grease, indicates that something is fat but may also be used in the sense we in English use the term "rich" about food. Usually it is used in a positive sense. In descriptions of fat food, such as pork or anything that has been prepared with coconut milk, it is not always clear without asking whether *gris* signifies that something tastes good or is just good, for example indicating that it is healthy. At the same time it is related to social status, an example being that those with the highest social status, such as big men/chiefs or white visitors/anthropologists are served the fattest food, such as the blubber from pigs. Whether food is regarded as good or not is sometimes described with another more general term, *gudfala*, which simply means good or nice, again ambiguously being unclear whether it means that something tastes good or is good for you. Finally there is the word pair of *strong/wik*, from the English strong and weak. I have rarely heard people use the term *wik*, which for example indicates that something does not taste of very much. It is almost, it seems, exclusively used in descriptions of liquid forms, such as too thin coconut milk, and would not be used to describe "tasteless" *nambo*, which would rather be described as *hemi no garem test* (it does not taste of anything). *Strong*, on the other hand, is often used in the sense we know from English when we, for example, describe coffee or tea as being strong but also used to describe that something is full of flavor, as in hot and spicy. In our early material it is hardly ever used, confirming that there was very little flavor in the form of spices and condiments added in food preparation. However, it also reflected the lack of a discourse on taste in general. This is something that changed mainly with the introduction of new food trends, to the extent that nowadays people often discuss food using the terms above, either in conversations with us, the visitors, in the context of hosting households, but also increasingly in public, such as when communal meals are served to e.g. people helping to build a house or on festive occasions. One of our main observations concerning food is thus that people simply speak more about food and taste than they did previously. We believe this to be the result of exposure to other food and taste cultures, either because Reef Islanders travel more than ever before or visits from people from outside, either from other parts of the country or from foreign countries, both known as *big ples*, or big place, in Pijin.

Fenualoa, one of the higher lying "main" Reef islands, has plenty of topsoil to sustain a substantial horticultural production, the gardens providing both an array of root crops and vegetables. There is an abundance of breadfruit trees, until recently[6] used to produce large amounts of *nambo*. Pigs and chicken provide the most commonly used meat, although pigs are usually reserved for special occasions such as funerals or age set rituals. Fishing, of fish and shellfish, also provides an important source of protein but is practiced less, and less efficiently, than on the Polynesian islands. I have never had a main meal on Fenualoa without tubers such as shade pana, sweet potato (*kumara*), very common although a relatively recent food crop,[7] cassava, believed to having been introduced in Polynesia in the nineteenth century, yams, as well as both cultivated and wild taro. Usually this would be combined in season with fresh breadfruit, or *nambo* out of season. To complete what may be described as a typical traditional everyday meal would be a sauce or stew made of more than fifteen different kinds of green leaves, all called *kabis* in Pijin—a word derived from the English cabbage—tomatoes and shallots. While this would be cooked in a pot over an open fire in the kitchen house, the other ingredients would either be boiled in a pot or cooked in an earth oven (*motu*), often "milked" with the ground flesh of a coconut. Until the 1990s it was very rare to add any flavor or condiment other than salt, which applies to all three locations.

Although one can wade across the reef to Nifiloli from the northern tip of Fenualoa at low tide, there is a significant change in the ecosystem, the island being a coral atoll with only a thin layer of topsoil. We are now among the people of the sea, which does not necessarily indicate a huge difference in food patterns but they rely much more on exchange in order to obtain root crops, of which only very few can grow, although there are quite a few wild taro. There are breadfruit trees but not as abundantly as on Fenualoa. Fishing is almost a daily activity, carried out by women and girls from the shore, both men and women from dugout canoes close to the village, or men on fishing expeditions to reefs in the lagoon or occasionally even beyond the reef, to the deep blue sea. Pigs are common here as well, mainly fed on coconuts, as are chickens. The ways of cooking are very similar to those on Fenualoa, i.e. mainly cooking in a pot over open fire or using an earth oven. Fruit bats and birds, often shot by young boys using slings, are supplementary protein sources.

Pileni is a real coral atoll, even smaller than Nifiloli and separated from the lagoon of the Reef Islands by a stretch of deep blue water. No root crops are grown here, so they have to be obtained through exchange, for example supplying fish and shellfish to the main Reef Islands, or even as far away as Nendö. There is a limited number of breadfruit trees, but in addition to plentiful coconut palms, there is a surprising number of cooking banana or plantain (*taveli* in Pileni) trees, explaining why this is the main difference in ingredients used in cooking, plantains served with most meals, if available combined with tubers. Although there are some pigs and chicken, fish and shellfish are eaten almost on a daily basis.

The three preceding paragraphs describe more or less what *kastom* or traditional food culture was in the Reef Islands for decades if not centuries, indicating the minor variations mainly due to availability of food resources. Apart from the main meals, usually served in the evening, with leftovers forming the breakfast of the following day, snacks are consumed either in conjunction with the meals, before or after, or in between. These consist of fruits such as mango, banana, Malay apple, pawpaw (papaya), and Oceanic lychee, and a wide variety of nuts, including different species of cut nuts and the very common ngali nut called canarium almonds (*canarium salomonense*) in English, eaten raw or roasted. Nutritionally speaking *kastom* food practices in the Reef Islands, as was the case in many other parts of the Solomon Islands, provided a very healthy and balanced diet.

By the time the Pinholts arrived in the early 1970s a few changes, which would change the taste of food, were gradually appearing. As is mostly the case with almost any kind of social change, this was mainly triggered by influences from the outside world. An increasing number of Reef islanders, initially mainly younger men, would migrate to find job opportunities and a cash income in other parts of the country, either by travelling to the national capital, Honiara, finding education and training, ending up as civil servants or working in the growing private sector, or finding jobs in either the oil palm plantations on Guadalcanal or the coconut (copra) plantations of Russell Islands, or in the huge fisheries plant in Noro in the Western Province or one of the many fishing vessels, belonging to Solomon Taiyo (now SolTuna), the joint tuna fishing venture of a Japanese company and the Solomon Islands government. The latter was the case for many young men from Vaiakau, men of the sea as they were. While these changes were brought about by Reef islanders going out, some changes were undoubtedly also linked to influences coming in, with ships coming to collect the rather limited number of goods that were exported from the Reef Islands such as pigs, copra and beche-de-mer (sea cucumbers).[8] A singular event that had lasting influence on life, and food, in the Reef Islands, was the arrival in the 1960s of a white couple, Tom and Diana Hepworth, who sailed into Mohawk Bay on their yacht, and fell in love with the place. After long and arduous negotiations with the big men they managed to obtain a ninety-nine-year lease of the tiny Pigeon (Ngarando) Island, which eventually became the most important copra buying point in the islands, the post office, a resort, and the most well-stocked store.

All these developments resulted in an exposure to the outside world that introduced many new goods, including food items, to the Reef Islands. Solrais rice, marketed as the Solomon Islands' own rice, had already for some years been imported from Australia, and was gradually becoming an important element in any meal, alongside tubers, breadfruit and plantains. By the time of my first field trip to Reef, in 1994, rice was no longer something only served on special occasions or once in a while, because you need cash to purchase it, but served with almost every meal, now with another relatively new food item, tinned fish (*tinfis*) from Solomon Taiyo. In pure economic terms it made little sense to pay for a tin of second-grade tuna when you could easily catch fresh fish just outside your home. Apart from being an easy and swift way to prepare a meal, boiling some rice and heating a few tins of tuna in brine or oil, many of my Reef Island friends quickly adopted the new taste it provided, and explained that it was almost a relief to the palate, not exactly using those words, after a diet of tuber-based meals, which they now began to criticize as being bland, or *hemi no garem test*, as referred to above. While I would never agree with them that second-grade Taiyo is particularly *suit*, a new lasting flavor had found its way into the Reef cuisine, although it was undoubtedly significant that food items bought with cash carried a certain level of status and prestige.

In the 1990s a number of decisive other developments took place in Reef food, all introducing completely new flavors, and becoming incredibly popular within a very short time span. Apart from the Hepworths' store, there were now small stores in most villages, all carrying the same items, which were all purchased through the Solomon Islands retail system, completely dominated by the Chinese Solomon Island minority in Honiara, perhaps the reason why most of these new elements were Asian. In 1995 I was invited to dinner at the house of Willie Maninga, a close friend and key informant, on Nifiloli. Willie had recently returned to the Reef Islands after spending years in Honiara, having been trained and worked as a policeman. Sitting on the ground on a mat woven of palm leaves, of which the women of Vaiakau are expert makers, traditional plates, made from banana leaves, loaded with boiled tubers and *nambo* (called *tupoe* in Pileni Polynesian) started to arrive. Just before saying grace, Willie had placed a pot, cheap Chinese aluminium kitchenware, on the mat, and during the short and simple prayer I sensed a waft of chicken combined with something I had never smelled before in the Reef Islands. The chicken stew was presented as both *suit* and *gris*, Willie evidently and nervously awaiting my reaction and verdict. It was absolutely delicious and I asked him how he had cooked it. Smiling he explained that had used *blak saus* (black sauce, which is cheap Chinese soy sauce), something a Chinese friend of his had introduced him to in Honiara. Not only was I witness to yet another new flavor introduced in the Reef Islands, but this is also one of the first recollections I have of a conversation centred on the taste of food, the emergence of a taste culture?

An almost identical experience took place the following year in Malapu village on Fenualoa, where Jens and I were guests at the house of James Temuka and his wife, Joanna Taipaki, arguably the best cook in the Reef Islands, which was acknowledged not only by us but by many people saying how good she was at preparing *suit kaikai* (delicious food).[9] We had often eaten there and Joanna knew that we liked a particular sauce she made with so-called slippery *kabis* as the main ingredient. I was sure that was what was in one of the pots placed in front of us but detected the distinct smell of something new. Joanna had made my first curry in the Reef Islands, explaining that "*hemi strong fogud*" (it is very strong, i.e. spicy). I did not find it particularly spicy, but the cheap, although not by local standards, Chinese curry powder she had used was definitely a completely new flavor. Around fifteen years later we stayed in the neighboring village of Tange for five days, where the village big men had arranged for two teenage girls to cook for us. Curry powder was used in every meal they served.

A more worrying new flavor trend, at least from a nutritional point of view, which started in the 1990s and was widespread after the turn of the century, was the introduction of artificially flavored noodles, or cup noodles as they are known. They have become extremely popular, and sometimes you see children eating them straight from the plastic container, without adding boiling water. But mostly they are used as a spice or as stock, adding the "powder" to stews and sauces or simply pouring a cup of noodles over the boiled tubers or rice. The Reef Islands, and their hunting (fishing)/gathering, combined with limited horticultural production, which only a few decades ago would come close to Sahlins' (1972) notion of stone age economics, have now truly met the globalized world of fast-food, and all the flavors that come with it.

Part of the above analysis may sound quite disquieting, especially from a nutritional perspective. Following our 2015 fieldwork, however, there seem to be more positive trends. Although we have not yet analysed all nutritional data, and conducted a comparative analysis with material we have from the 1970s and 1990s, we seem to be able to outline three distinct periods, each with specific food trends. First, the 1970s, during which the Reef Island food culture by and large seems to fit what we may describe as *kastom*, i.e. as it had been for ages, almost exclusively based on local produce, although imported rice was becoming increasingly common. Secondly the 1990s, where we see a rapid increase in the consumption of imported food items, rice becoming a staple food often replacing local root crops or breadfruit and plantains, and tinned fish becoming almost more commonly used than fresh fish. White bread is also becoming common, more frequent travel to the provincial capital of Lata or the national capital of Honiara initially giving access to bakeries, later importing white flour and dry yeast and baking it in the villages. When it comes to new flavors, as we have seen, this is also when, for example soy sauce, curry powder, and eventually flavored noodles are introduced. Finally, the current decade, if our material from 2015 is representative, seems to indicate a return to the roots, if readers will pardon the pun. There is a return to local fresh produce, with gardening activities expanding on Fenualoa and fishing becoming the normal thing to do rather than buy tinned fish. The main threat to Reef food culture today seems to be not so much the influx of imported food that we saw only a few years ago, but more the impacts climate change and shifting weather patterns may have on the productivity and seasonal harvesting of—more important than any other crop—breadfruit.

If we define the source of taste or the act of tasting, what happens when we put food or drink in our mouth, as culture and nature becoming one, as Trubek has it (2008: 6), the main changes to Reef taste in the future may thus be triggered by changes in nature—not only changing weather patterns but perhaps even more seriously rising sea levels, which already threaten literally to wash out the low-lying coral atolls of Vaiakau. What I have hoped to show in this chapter, however, is how various cultural factors, most notably increased exposure to the outside world, have fundamentally changed Reef food tastes, possibly even giving birth to a new taste culture that regards food as something that moves beyond plain sustenance. The most culturally iconic food item of Reef, *nambo*, may still in itself remain almost tasteless, but the culinary context in which it is now served has undergone radical changes over a period of thirty to forty years. Now there is a local craving for Reef food that is both *suit*, *gris*, and *strong*.

Notes

1 The Danish International Development Agency.
2 Almost needless to mention this chapter is dealing with taste, and flavor, in the specific context of food, and to a certain extent drink, while acknowledging that other aspects of taste do exist, not only as metaphors or expressions of food culture, such as tastes of fashion or music, or non-gastronomical usages of terms such as "delicious" or even "dishy."
3 My colleague, and co-editor of this volume, Susanne Højlund, has recently (2016) published an excellent overview booklet on "taste" in Danish (*Smag*). Carolyn Korsmeyer some years ago edited a textbook, *The Taste Culture Reader*, which is now out in its second edition (2017), containing important excerpts from works in the field, going back to the early seminal work of the gastronomer, and lawyer, Jean-Anthelme Brillat-Savarin (1755–1826).
4 This cultural culinary significance is akin to Elisa J. Sobo's (1997: 258) analysis of Jamaican food culture in which food is unthinkable without tubers—food actually meaning tubers.
5 While Pileni Polynesian resembles many other Polynesian dialects many words are understandable if you have some rudimentary knowledge of Polynesian. Aiwoo, on the other hand, is regarded by linguists as one of the most difficult languages in the world. Two Norwegian linguists are among the few to have worked with both languages (see Næss 2011 and Næss and Hovdhaugen 2011).
6 Although it has not yet been scientifically validated climate change and changed weather patterns are seriously affecting the "seasons." There used to be two harvest seasons for breadfruit, a small one in February/March, during which most breadfruit were consumed fresh, and the main harvest in August, during which the production of *nambo* took place, producing sufficient to even export large amounts to the capital, Honiara, bringing in cash, increasingly needed to purchase foreign goods as well as pay school fees.
7 Originating from the "new world" of tropical America but known to have been introduced to the Melanesian part of the Pacific as early as the seventeenth century (Massal and Barrau 1956).
8 By this time one item that had previously been a main source of income, especially for the small coral atoll populations, were shells, used in the garment industry worldwide for making buttons, now replaced by the much cheaper material: plastic.
9 I believe Joanna's food stands out because her recipes are more complex, using more ingredients than most others, creating distinct flavors even before she had access to imported spices and sauces. I have over the years noticed one interesting trait when observing her cooking, namely that she actually tastes and samples the food before serving it, something I have rarely seen others do.

References

Arnott, M.L., ed. (1975). *Gastronomy. The Anthropology of Food and Food Habits*. The Hague: Mouton Publishers.

Brillat-Savarin, J.-A. (2017 [1825]). "On Taste," in Korsmeyer, C., ed., *The Taste Culture Reader. Experiencing Food and Drink*, 2nd ed., 13–20. London: Bloomsbury Academic.

Cowling, C. (2008). Temotu Nutrition and Food Security Project. Baseline Survey, Honiara: World Vision.

Crawford, P.I. and J. Pinholt (2010). *Solomon Islands is our Country—Reef Islands is our Home*. Højbjerg: Intervention Press.

Davenport, W. (1969). "Social Organization Notes on the Northern Santa Cruz Islands: The Main Reef Islands." Baessler-Archiv, Neue Folge, XVII: 151–243.

Goody, J. (1982). *Cooking, Cuisine and Class: A Study in Comparative Sociology*. Cambridge: Cambridge University Press.

Højlund, S. (2016). *Smag*. Aarhus: Aarhus University Press.

Koch, G. (1971). *Materielle Kultur der Santa Cruz-Inseln unter besonderer Berücksichtigung der Riff-Inseln*. Berlin: Museum für Völkerkunde.

Korsmeyer, C., ed. (2017). *The Taste Culture Reader. Experiencing Food and Drink*. 2nd ed., London: Bloomsbury Academic.

Macbeth, H. and S. Lawry (1997). "Food Preferences and Taste: An Introduction," in Macbeth, H. and S. Lawry, eds., *Food Preferences and Taste. Continuity and Change*, 1–113. Oxford: Berghahn Books.

Massal, E. and J. Barrau (1956). Food Plants of the South Sea Islands, South Pacific Commission Technical Paper no. 94. Noumea: SPC.

Næss, Å. (2011). "Cutting and Breaking in Äiwoo: Event Integration and the Complexity of Lexical Expressions." *Cognitive Linguistics*, 23(2): 395–420.

Næss, Å. and E. Hovdhaugen (2011). *A Grammar of Vaeakau-Taumako*. Berlin: de Gruyter Mouton.

Norbye, A.-K. Brun (2010). "Eating Memories: A Taste of Place," in S. Williksen and N. Rapport, eds., *Reveries of Home. Nostalgia, Authenticity and the Performance of Place*, 145–163. Newcastle-upon-Tyne: Cambridge Scholars Publishing.

Rozin, E. and P. Rozin (2017). "Culinary Themes and Variations," in C. Korsmeyer, ed., *The Taste Culture Reader. Experiencing Food and Drink*, 2nd ed., 37–43. London: Bloomsbury Academic.

Sahlins, M. (1972). *Stone Age Economics*. Chicago: Adije-Atherton.

Sobo, E.J. (1997). "The Sweetness of Fat. Health, Procreation and Sociability in Rural Jamaica," in C. Counihan and P. Van Esterik, eds., *Food and Culture. A Reader*, 256–271. New York: Routledge.

Sutton, D.E. (2010). "Food and the Senses." *Annual Review of Anthropology*, 39: 209–223.

Sutton, D.E. (2017). "Synesthesia, Memory, and the Taste of Home," in C. Korsmeyer, ed., *The Taste Culture Reader. Experiencing Food and Drink*, 2nd ed., 303–314. London: Bloomsbury Academic.

Trubek, A.B. (2008). *The Taste of Place. A Cultural Journey into Terroir*. Berkeley: University of California Press.

Tuan, Y.-F. (2017). "Pleasures of the Proximate Senses: Eating, Tasting and Culture," in C. Korsmeyer, ed., *The Taste Culture Reader. Experiencing Food and Drink*, 2nd ed., 221–228. London: Bloomsbury Academic.

Yen, D.E. (1975). "Indigenous Food Processing in Oceania," in M.L. Arnott, ed., *Gastronomy. The Anthropology of Food and Food Habits*, 147–168. The Hague: Mouton Publishers.

Part Two

Taste, Place, and Intersubjectivity

5

How to Taste Like a Cow

Cultivating Shared Sense in Wisconsin Dairy Worlds

Katy Overstreet[1]

Introduction

After the hay sale, Margie, her son Steve, and I sat down to drink tall frosty glasses of rootbeer. The weekly auction-style sale takes place in the parking lot of a gas station in rural southern Wisconsin. Half the gas station convenience store is dedicated to a fast food restaurant, which offers sodas, hot dogs, and hamburgers. Margie and Steve attend the hay sale every week to scout for bargains on hay bales for their approximately sixty-cow dairy herd. Margie gossiped with two of the auctioneers at the table next to ours as they ate hot dogs. Steve and I continued our conversation about how he chooses particular bales of hay to buy at the auction. Steve leans forward and says, "The best way to do that is find something that you would want to eat if you were a cow. You gotta think like a cow. Is this something you're gonna want to eat? Because if you're going to be willing to eat it then the cows will eat it. That's the way I do it."

Before the hay sale each week, Steve and the other buyers walk through the gas station parking lot, carefully assessing the qualities of the various hay bales. Several times I followed along with Steve as he examined the hay by sight, smell, touch, taste, and more. Through sensory cues he evaluated hays for nutritional value, freshness, digestibility, and palatability. While Steve references formalized knowledge such as research documents publicly available through University Extension as he considers nutritional values, Steve claims to assess palatability through "thinking like a cow." But what exactly does he mean by this? Certainly there is a sense of "cow" as a sentient and perhaps thinking "other" in Steve's description. But when Steve told me to "think like a cow," we were discussing feed specifically, rather than the general cognitive abilities of cows. Therefore, Steve means something along the lines of "thinking through the senses" (Classen 1993) and thus I will pose the question instead as: what does it mean to taste like a cow?

While many studies of animal agriculture argue that animal others are rendered as objects through their use for producing profit, I found through research among Wisconsin dairy farmers that even on farms of 500 or more cows, commodification was never complete; it was one among many ways of relating between humans and cows. Even during routine handling or painful interventions, cows were still understood to be subjects and on most dairy farm visits I found that at least some cows were known as individuals, pets, or family members. Over the course of almost two years of fieldwork with dairy farmers and dairy experts in primarily south-central Wisconsin, I participated in milking and chores on five farms and visited over 100 farms of varying sizes. I also rode along with experts important to many aspects of cow care including veterinarians, dairy nutritionists, extension agents, and siring specialists. I came to see how the "ontological choreography" or the coordination of various ontological orders legal, medical, religious, and otherwise (Thompson 2005) through which cows and humans are kept separate, at times becomes very messy. In other words, it was not always clear where cows ended and humans began. Traditionally, phenomenological and social science accounts of sensory experience, while at times engaging questions of intersubjectivity, or the social relations between beings that recognize each other as subjects, continue to confine sensory experience to individual bodies even while they show that senses are shaped through cultural values and culturally informed "modes of attention" (see for example Csordas 1993 and Grasseni 2009). Through examination of how farmers, dairy nutritionists, and cows select feed, I aim to show how senses, in this case a sense of taste, might be understood as something that transcends the assumed edges of individual species-bodies. As Steve tastes hay or as dairy nutritionists sample feed, they are working to taste not for *themselves* but for *others*. And thus they extend their sensorium towards that of cows. They do so through sedimented experience and skilled practice.

At the same time, cows, too, learn to taste. Cows on dairy farms in Wisconsin, also known as America's Dairyland, commonly eat substances that might have been unimaginable as a food for cows 100 years ago. Further, their diets are based heavily on grain in order to support high levels of milk production. In other words, they too have learned to eat cow feed. Therefore, the taste for cow feed is not-quite-human and not-quite-cow. It is heterogenous and recursive. Taste emerges recursively as humans observe what and how cows eat while cows eat feeds shaped by available industrial byproducts, ideas about the productive modern dairy cow, and local proclivities towards particular flavors. Through the circling kinds of eating and observations of others eating, taste moves through and across bodies: it is *transcorporeal*. Through an examination of how taste for cow feed emerges through practices of care, noticing, biopolitical control, and resistance, I aim to push for a sensory model that moves beyond phenomenological preoccupations with separable individual bodies as the base unit for sensory experience.

Cow Sense

As Steve walks through the hay sale, he attends to the qualities of hay bales through multiple forms of sensing. He examines hay bales visually before he approaches them, which enables him to skip a further assessment if he finds the bales on a particular trailer to be undesirable. Indeed the qualities of the bales are various although this is not always obvious from afar. The trailers are a hodge-podge ranging from those that lean and creak, revealing their age through rusty rails and cracked wood to ones that look almost new. The hay bales themselves range from large cylindrical "round bales" held together with plastic sheeting or twine to cuboid "square bales." Some bales are tightly packed while others seem to resemble almost a pile. They come in shades of yellow, green, and brown, some dry and pale from sun exposure and others holding richer colors.

As I tagged along with Steve, he walked right past hay that to my untrained eye looked just like the previous batch. When I asked Steve why these bales did not appeal to him, he led me closer to the bales and pulled a handful out of the middle of the large round bale. "This is first crop hay here," he told me as he held the sample out to me. Farmers often cut "hay," which is a mixture of grasses and forbs, several times throughout the growing season. Steve pointed out that the stems of first crop hay are often much thicker than the stems in later cuttings and I could easily see that, compared to the last bale of "third crop hay" or "third cutting", these stems were much sturdier and thicker. Steve said that while he is not opposed to feeding his cows first crop hay, it can be too "pokey" for them, making the hay less palatable and more likely to cause jaw abscesses. It would certainly be too coarse for the sensitive mouths of his young heifers.

As he described the difficult texture of first crop hay, he poked the tips of some of the stems into his hand as if to emphasize the way he, too, found the texture unpleasant and that his hands could be sensitive to the coarse stems in a way that might tell him something about the experience that cows might have when eating the hay. His hands could tolerate the rough stems and at the same time discern that the degree of coarseness would be uncomfortable and potentially even harmful for the sensitive mouth interiors (palate, gums, and cheeks) of his cows and the young heifers in particular. His hands thus imagine themselves to be the chewing mouths of cows as they rub the hay sample between the fingers. They discern the softer textures of later crops of hay as something more palatable to Steve's heifers. In this way, Steve's hands might be understood to "taste forward" or anticipate the potential eating experience of a bovine other. Steve's use of his hands to anticipate the palatability of hay for his family's dairy cows and heifers raises an important question of where taste happens.

In an experiment aimed at testing the remark of one participant that food tastes better when eaten with the hands, Mann et al. (2011) prepared and ate a dinner with a variety of flavors and textures, which they consumed using their fingers and hands. As they mashed, rolled, and mixed food before bringing it to the mouth, they found that their fingers and hands anticipated how food might feel in the mouth in terms of texture and flavor and thus participated in the sensory experience of taste (Mann et al. 2011; see also Janeja 2010). In a similar way, Steve's hands take part in the tasting of hay and the anticipation of mouth feel. Thus taste extends beyond the mouth and the moment of tasting is initiated before food enters the mouth.

Continuing our survey of the hay bales that would be auctioned off, Steve and I approached some cuboid bales. He used these bales, which were loosely held together by twine to show me that sometimes a bale can appear to be in fine shape but that the interior can hide something different. "You want to make sure that the bales are tight," Steve said. "If they aren't tight then they fall apart when you move 'em around. And they might go bad inside too." He pulled out a handful of hay and smelled it. "Put your hand

Figure 5.1 Hay sale: A farmer smells the hay at the weekly hay sale. Photo by Katy Overstreet.

in there," he said to me. The hay bale was wet on the inside despite the dry appearance on the outside. Then he held out the sample for me to smell. "Smell that sweaty sweet smell? That means it's too wet. Yeah that'll have to get fed up right away," Steve said as he stuffed the sample back into the bale. We moved on to the next trailer.

Steve had purchased hay that was too wet before and he said that it had been a waste of money because they had not been able to use it all before it molded. Cows, he told me, will tolerate a little bit of mold but it could make them sick. Thus, as Steve assesses the hay at the hay sale, he draws on memories of sick cows as well as local commonly held knowledge of what could make cows sick, in addition to information distributed by the local University Extension agents about the possible perils of hay such as nitrate poisoning, mold, or inadequate nutritional content for lactating cows in particular. As Lévi-Strauss showed, Steve's discernment between rotten and fresh feed is shaped through shared cultural understandings of these qualities as well (1997).

Steve brings his body, through the work of his senses to the task of selecting feed. This is what he means by, "You gotta think like a cow." Steve is thinking like a cow by sensing according to a cow-ness as he has come to understand it through working with many different cows. Much of what he's learned has been from being with his family's cows as they eat, as they get sick, and so forth. And so he has become sensitive to the bodily needs and vulnerabilities of cows in different life stages. Steve has trained his senses towards an attunement with what constitutes palatability through aroma, texture, and flavor for a cow. In other words he has developed a skilled sensorium (Grasseni 2007, 2009) by honing his attention to particular discernible characteristics of feed and through bringing multiple senses to the anticipation of taste. David Sutton describes the ways that senses come together as synesthesia (2001, 2010) inspired by the tendency of Kalymnian islanders to describe the sound of a smell. Steve's sensory assessment of the hay is thus constituted through memory, through synesthetic attention to and experience with cows, and through institutional knowledge.

Through feeding cows, Steve and his family seek to provide sustenance as well as to prevent harm through sickness or injury. Feeding cows is a form of care at the same time as it is a necessary part of milk production. In order to prevent harm, Steve draws on transcorporeal taste, or that which is trained through experience being with cows in order to assess everything from aroma to texture to the digestive effects that different feeds may have. For instance we examined a trailer piled high with alfalfa hay and Steve thought that while this might be good in a mix, it might give his cows diarrhea if fed alone. As farmers and nutritionists consider digestive effects alongside palatability, they demonstrate how cultural studies of taste have perhaps been too limited in considering only the sensations of the mouth rather than what happens throughout the gut (Caldwell 2014). This is a key aspect of transcorporeal taste: humans who develop cow sense are able to discern the potential for particular digestive effects of feed for cow others even while human and cow digestion differ significantly. While the qualities of the hay may not effect Steve's body in the same way as that of cows, his senses extend towards potential futures of eating and digesting of bovine others.

Margie and Steve have a small farm and are responsible for all the feeding of their cows through a combination of growing feed crops, grazing cows in their wood lot, and purchasing feed through hay sales and feed dealers. They cobble together different sources of feed in order to make an affordable and nutritious diet and they remain "self-taught" in that they draw on their own experience, accrued family knowledge through multiple generations of dairy farming, vernacular knowledge through talking with neighbors and friends, and their own research into formalized knowledge through publicly available resources largely through University Extension. Margie and Steve's strategy is exemplary of that of many smaller-scale farms in the area. On the other hand, larger farms have come to more heavily rely on professional dairy nutritionists to design diets and evaluate the nutritional content and the costs of different feeds and supplements.

During my field research in Wisconsin, I interviewed and followed several dairy nutritionists as they evaluated cattle and feed on farms and designed diets. Through this research, I saw that dairy nutritionists, like farmers, cobbled together information in order to design diets. They were more likely to rely on formalized knowledge in the sense that they followed university-based research on dairy nutrition, they sometimes had a Masters degree or other kind of certification in dairy nutrition and they also drew on information on feed and cows from their companies be they private corporations or member-owned cooperatives. While these sources figured in significant ways that shaped the diets they produced, dairy nutritionists also drew on transcorporeal taste in ways similar to farmers. I did not meet a single dairy nutritionist in my time in the field who had not spent significant time with dairy cows while growing up. I met one who did not grow up on a farm, but she had spent a great deal of her childhood on a neighbor's farm. She worked on this dairy farm throughout high school and college and is now their dairy nutritionist. In Mathew Desmond's study on wildland firefighters, he examines how many of the firefighters embody a self-described "country boy" *habitus*, or embodied way of being in the world, which translates well into their work as firefighters (Desmond 2006). In a similar way, the dairy nutrition profession is largely made up of people who have developed cow sense or a sensory attunement to cow ways of being. And so, the kind of sedimented sensory experience that enables farmers to imagine how feed might taste to a cow, also figures in the work of more formalized nutritional planning. Consider the following example.

While attending the World Dairy Expo (WDE) in 2014, I met an animal nutritionist who worked specifically with dairy farms that use robots to milk their cows. Many milking robot companies and owners claim that cows like robots because they can decide when to get milked. It is common practice to provide an "added incentive" for the cows to come to the robots in the form of pellets that drop down for the cow when she is getting milked. Brian, a dairy nutritionist who works specifically with robot farms, spoke at WDE about how he designs pellets for robot farms. He said that he designs pellets based on what cows like to taste by tasting them himself. They need to taste good to him in order for the cows to find them palatable as well. Later, during a phone interview, I asked Brian how he knows what will taste good to a cow.

"Very simple. Back when I was a kid, and we were feeding roasted beans, we kids would eat them too. They sort of taste like a peanut. So you develop taste." In addition to tasting the feed that cows are eating, Brian described watching what cows eat and what they seem to prefer. He described how sometimes developing pellets required trial-and-error in order to see what cows like best. Taste, then, for dairy nutritionists is something separate from that which is learned through formalized courses in dairy nutrition or scientific research. It is something accrued through long-term relations with cows on dairy farms as well as careful observations of cow preferences. In order to develop taste, humans must pay attention to what cows will eat and how they eat.

Nutritionists often described to me how important it was to "listen to cows." John, a dairy nutritionist who generously let me ride along several times with him as he visited farms and assessed the feed and the diet-related condition of cows, told me that the "cows will tell you what you need to know" if only one pays attention. He told me stories of visiting farms where he saw cows drinking urine or "consuming hillsides" (i.e. eating soil), which he read as a sign that the cows' diets were deficient in important minerals. Lyndsey, a farmer that I worked with frequently throughout my fieldwork, has one cow named Iodine because they give her a squirt of iodine in her nose and mouth every feeding, a practice that she started after Iodine regularly sucked on "cow towels" leftover from cleaning teats with iodine. Now Iodine "asks" for iodine at each milking through turning back to the person milking her, and licking them until she gets her squirt. Farmers and nutritionists also try to watch out for how the social dynamics among cows affect eating. Some cows tend to "boss" others by pushing them away from feed. Farmers and nutritionists pay attention to these types of behaviors and sometimes address the issues that arise for particular cows through grouping cows differently or changing how they offer feed.

Transcorporeal taste emerges through these careful practices of noticing and copresence, a term that Barbara Smuts uses to describe the experience of singular awareness with another, and in particular a member of another species (2001). Barbara Smuts and Donna Haraway both describe experiences of "deep intersubjectivity" that have arisen out of long-term relationship and being with their dogs. As Barbara Smuts argues, this shared awareness can only arise through "pureness of motives" (2001: 308) meaning not that the relationship is free from all utilitarian benefits but that the experience "cannot be coerced but must, by definition, reflect independent agency by each animal" (308). It would seem that Smuts' definition of this shared awareness would preclude the relationships of humans and livestock animals in production agriculture, even while she does not seem to fully account for the disparity in power between her and her dog, Safi. It would certainly be erroneous to describe dairy cows as having independent agency when they are subject to deep forms of biopolitical control on a daily basis not to mention frequent and sometimes painful bodily modifications. Yet even these forms of control open possibilities for being with, such as the kinds of "shared suffering" that workers and pigs experience through the course of working together on industrial pig farms (Porcher 2011).

Donna Haraway makes theoretical space for deep intersubjectivity in spaces where animals are involved in relations of use that are highly controlled by humans and that may involve suffering or death on the part of the animals (2008; also see Hayward 2010). In the case of dairy cows and farmers, the unequal power relations mean that not everyone comes to the relationship entirely by choice. Cows do not decide to work in dairy. Yet, dairy farming relies on the kinds of everyday willingness of cows to participate in work (Porcher 2012) which becomes unavoidably clear when cows refuse to do what is expected of them as workers (Porcher 2012; Despret 2016). I would argue that transcorporeal taste speaks to a depth and a kind of shared bodily attunement that is related to Smuts' conception of copresence. As farmers and nutritionists learn to taste like cows, they are shaping their awareness towards that of another. And this awareness, or what we could call "cow sense" only becomes possible through long-term being with. This is not to say that every person who worked frequently with cows that I encountered in the field had what I am describing as "cow sense."

On one of the farms I frequented throughout my research, Diane and Bill Brown split the farm work; Bill did the cropping while Diane managed the cows. Bill grew up on a dairy farm and generally liked the work but he would easily get frustrated with the cows. He would shout in the milking barn or be quick to strike them on their flanks or to twist their tails. Diane told me that the cows were always jumpier when she returned from a trip because of her husband's impatience with them. Thus, cow sense also requires an interest in noticing cows that not all dairy people engage.

As numerous social science scholars have shown, food often becomes a medium through which social relations are enacted. Foods and food sharing create and sustain social networks. Foods inflect memories. Foods are political and laden with meaning and histories. Like Mauss' description of gifts, food can be thought of as a "total social fact" (Mauss 2000). Generally anthropological accounts of food as a social medium are confined to sharing between humans. In this case, the consumers are cows. But this does not mean that acts of feeding become devoid of social meaning for either humans or cows. Rather, farmers describe feeding their cows as a form of "care" often inflecting both their sense of responsibility to the cows from whom they make a living and their felt affection for their cows. Feeding cows might involve empathetic and compassionate registers, but it is also done in the name of production. Cows are fed rations that are geared towards producing the most milk in the smallest amount of time at the least cost. Feeding for production can lead to many issues for cows including acidosis, laminitis, and liver abscesses. The feeds and practices of feeding in dairy are also shaped through trends in nutritional science, seed technologies, milk prices, state and federal legislation, consumer desires, and more.

Feed therefore takes on many meanings at once. It becomes a technology for milk production as well as a form of intimacy and care between farmer and cow. Feed and feeding practices can also be a marker of a farmer's skill as a manager. Many times I heard farmers or veterinarians remark on the skill of a farmer based on the state of their cows. If the cows were deemed too skinny then the farmer was by implication a poor manager. Cows with luster to their coats and who come through the winter without losing too much weight are described as a sign that a farmer "knows what they're doing." In this way the ability to "taste like a cow" matters for both human–cow as well as human–human relations.

Cultivating Taste

Feeding cows on dairy farms involves affective registers of responsibility and nurturing through food provision. Yet the relations between humans and cows in dairy barns are oriented towards production and thus are not necessarily about providing the kinds of foods that are the most health-giving for cattle. Instead these diets are about increasing milk production and sustaining cow bodies enough to carry on in the work of producing milk, an attitude towards animal welfare in livestock agriculture that Donna Haraway describes as "just healthy enough" (2008). While most farmers grow at least some portion of their feed, feed costs average around 30 percent of input costs for dairy farmers and have at times threatened to put farmers out of business. For example, the drought of 2012–2013, largely through the spike in feed prices alongside crop failures, meant that many dairy farmers "paid to produce milk" as they put it. Most farms survived the drought but many did not. Feed, while a source of risk, is also a key means through which dairy farmers have sought to improve their herd productivity. It is considered common sense in the dairy industry that improved feeding practices have enabled the unprecedented levels of milk production per cow that we see today. Dairy farmers seek to make a profit, which motivates farmers to find ways to meet particular energy and nutritional requirements through inexpensive materials. And this requires training cows to develop tastes for particular kinds of feedstuffs and using products to make what would normally be unpalatable to cows more appetizing.

During my fieldwork, I spent time milking and doing chores on five different farms. Usually I was given "calf chores," which involves weaning calves from milk to solid feed. There is of course a lot of variation in caring for calves but it is possible to describe a pattern based on my experience doing a lot of calf chores and observing calf chores on many farms. During the first days of their lives, calves are given colostrum from their mother or another cow who recently "freshened" or calved. Then calves are switched to bottles of milk or a sweet-tasting "milk replacer." Calves then learn to drink from buckets. I learned to train a calf to drink from a bucket by letting the calf suck my

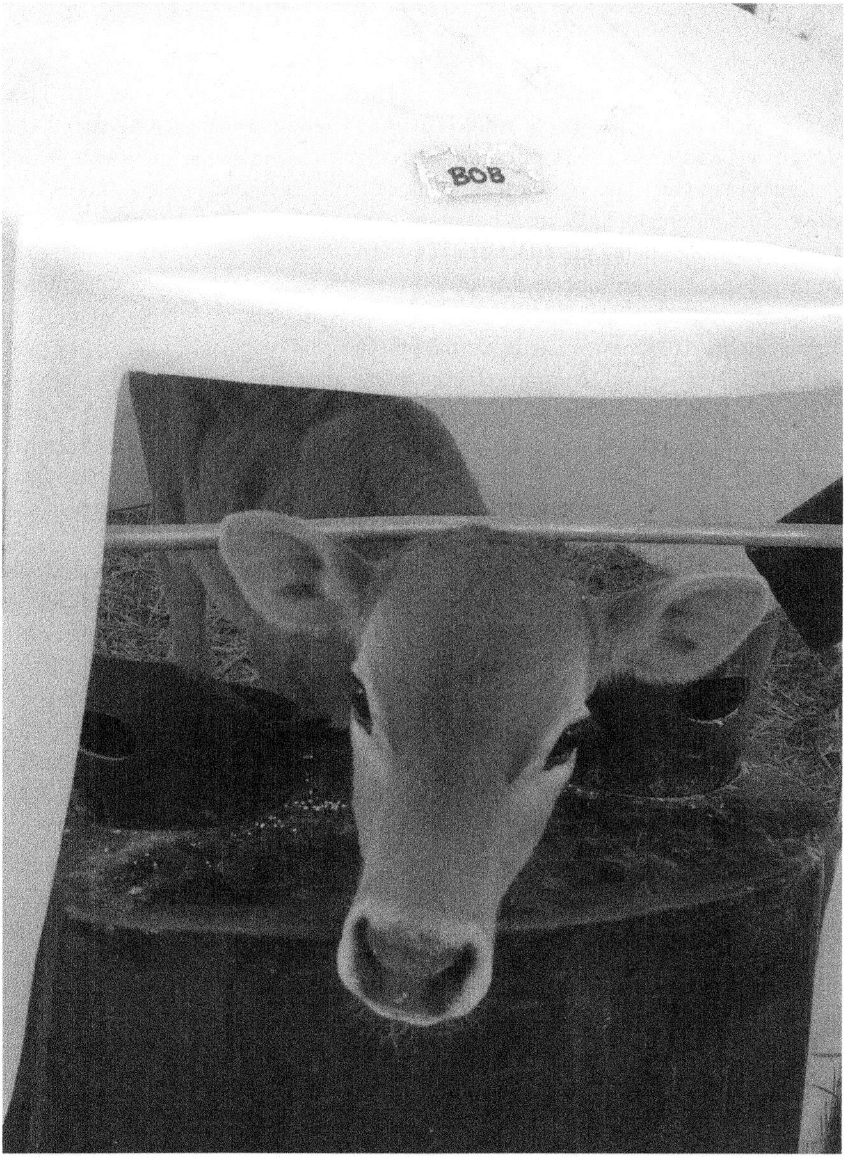

Figure 5.2 Calf: A calf with buckets of water and calf starter. Photo by Katy Overstreet.

fingers while slowly pulling my hand down into a bucket of milk and then repeating this until the calf learns to slurp up the milk. After the calves finish their buckets of milk, I was taught to mix a bit of "calf starter" into the remaining milk at the bottom of the bucket and then to offer that to the calves. Additionally, between milk feedings, calves are usually left with a bucket of water and a bucket of calf starter with the assumption that they will eventually figure out how to eat and drink them on their own. Calf starter is usually grain-based pellets or rolled feed. Calves must learn to eat grain because this is the basis for a production-oriented diet.

Since the 1950s, American dairy farmers have largely switched to a feeding technology known as a Total Mixed Ration (TMR). TMR has gained a lot of traction as a means to standardize dairy cow diets in order to allow farmers and nutritionists to design diets according to nutritional science recommendations for amounts and percentages of particular vitamins, minerals, carbohydrates, proteins, and more. A key driver for adoption of TMR diets has been the desire to incorporate cheap industrial byproducts from milling, manufacturing, and agriculture into cattle feed. In order to do so, these byproducts are measured according to characteristics such as digestible protein or vitamin content. Through evaluating byproducts as inputs of protein, vitamins, or carbohydrates, cows have come to eat the otherwise unimaginable; today dairy cattle on TMR diets frequently consume chicken litter, cottonseed, candy and chocolate, fish meal, and pork blood. These foods are only made appropriate as cattle feed through translation (Callon 1984) into nutritional components. Such incorporation of industrial byproduct is not new. In the 1980s and 1990s researchers attributed an epidemic of bovine spongiform encephalopathy or mad cow disease to the incorporation of cow bone and meat meal, the byproducts of slaughterhouses, into cattle feed.

A TMR is designed to be a complete balanced diet for cattle. And it is designed with the idea that every bite will provide balanced nutrition. In order to achieve this, TMRs are blended to a homogenous consistency from various materials. Depending on the farm, either the farmers or their nutritionists will design TMR diets through assessing the nutritional content of crops and stored feeds and through adding supplements to help the diet meet industry standards for nutritional content. Some farms mix their own TMR while others purchase mixed feed through dairy service cooperatives. It is worth mentioning that Steve and his mother Margie do not use a TMR. Like many small farms, they do not have the financial capital that it requires to either pay for mixed feed or to purchase the equipment necessary to make their own. Therefore it is a technology that is more accessible for farmers who are able to take advantage of economies of scale.

As I rode around with animal nutritionists and discussed feed with farmers, a frequent concern about TMR feeding emerged: feed sorting. TMR diets are designed to deliver an optimally balanced nutritional package in every bite. But cows might not always agree with this optimized kind of eating. Instead, the cows often push the feed around with their noses in order to eat the pieces of feed that they prefer, undermining the careful design of the blend by farmers and nutritionists. This places the humans in a bit of a dilemma: blending and chopping the TMR too much will reduce particle size which can lead to digestive issues for cows while too little blending enables cows to choose what to eat, which can mean deficiencies or overeating of feed components.

Figure 5.3 Feed: A dairy nutritionist examines a TMR for particle size. Photo by Katy Overstreet.

When cows sort for too much grain or components, they risk acidosis or other metabolic and digestive issues. When they sort for too much forage, they may not make the energy requirements of high-milk production.

In addition to problems with feed sorting, cows may refuse feed altogether. Angela, a dairy nutritionist that I rode along with several times, described how cows often refuse TMR feed when diets are not properly designed to mask the bitter flavors of added vitamins and minerals or when industrial byproducts vary in flavor or texture. As she designs diets for clients, she often includes feed flavorings in order to make the feed more palatable and to limit stress for cows when feed components change by maintaining aroma and flavor in the feed. Angela showed me a brochure from a company called QualiTech™ for dairy feed flavor enhancers with names such as "Caramel Delight," "Accelerate Berry," and "Butterscotch Savour". At the time, I was surprised that cattle feed enhancers sounded more like ice cream or candy flavors. But these flavors point to similarities in cultural perceptions around food provisioning as well as localized tastes for sweet foods.

A major challenge for many anthropologists is the adjustment to local eating practices. Before fieldwork, I spent many years participating in the "health food" worlds of California through working at farms, non-profit organizations, and cafés. In these eating arenas, sugar is not entirely absent but there is a propensity towards limiting or completely abstaining from processed white sugars. Like most Wisconsinites, my hosts had strong proclivities towards sweet foods, often purchasing ready-made sweets and baking desserts in the evenings and on weekends. Unlike California health worlds' vilification of sugar and sometimes sweets more generally, my host families considered sweets to be part of a healthy diet as long as they are consumed in moderation. To my sensibilities, Wisconsin foods were ubiquitously sweet. When offered a glass of wine for example, the options might be "chocolate wine" or "cherry wine" both of which would be syrupy sweet. Dinners at the houses of acquaintances and friends in Wisconsin might include such side salads as Jell-O or whipped cream salad as part of the dinner plate. Restaurants, gas stations, and farmers usually offered coffee accompanied by sweetened and flavored creamers. While living with Barbara, a nurse who grew up on a farm in the area, I once had a visitor who stayed with us who did not enjoy sweet foods. Before my friend's arrival, Barbara had generously baked a loaf of banana bread, a tray of chocolate bars, and another of Rice Krispies treats™. This was a common activity for Barbara who usually baked at least one sweet treat but usually more on her days off. My visitor left without partaking of any of Barbara's home-baked goods. Barbara was upset and tearfully explained to me the rejection she felt after providing these foods that went un-eaten by my guest and myself. "Food is how I show love," she told me after my guest left. It is not surprising that food and food provisioning is imbued with affect and the kinds of relationship building spirit that giving material goods entails (Mauss 2000). What is interesting here, however, is that this sense of caring through food is extended to cows and that cultural proclivities toward sweetness (Mintz 1996) inform efforts to select feed and to make cattle feed palatable.

Sweetness and sweet foods are understood to be a basic pleasurable good. It is perhaps unsurprising then that in an effort to make cattle feed more palatable, farmers and nutritionists would do so through making the feed sweeter and more like a fruit or dessert. Farmers and nutritionists often told me that cows enjoy sweet flavors, much like humans. For example, Jessica, a young farmer who had recently taken over her parents' farm, described how her cows sort through the hay that Jessica's husband cuts from their field: "I've seen cows go through and pick out the clover flowers. It's sweet. Cows would go through and pick out all these flowers and then they'd go through all the rest. It tastes just like honey. I like it. So . . . cows'll like it. I've seen them do it." There is an assumption here that cows prefer sweet foods. But rather than refer to some kind of predetermined biological bovine taste preference, I would argue that dairy cows, like their humans, are cultural eaters. They learn as calves to eat grains and calf starter, which usually contain molasses. Therefore it seems unsurprising that they might often sort for these kinds of foods or prefer sweetened feeds.

Wisconsin dairy cows then are learning to eat foods that have cultural relevance and value as "good" in the sense that sweetness is understood to be a basic pleasure in life. But, like humans, cows also prefer some flavors over others, and they do not necessarily eat in standardized ways. Despite the dream of a standardized eating that imbues the TMR philosophy of feeding cattle, on a daily basis cows are frequently resisting an approach that treats them as homogenous eaters. If Steve and Margie's abilities to anticipate the tastes of their cows constitutes transcorporeal taste wherein both cow and human senses are shaped through close intersubjective encounters, then cows refusing to eat TMR mixtures or sorting the feed constitute the failures of anticipatory taste. As opposed to empathetic encounters wherein understanding occurs, this resembles what Douglas Hollan (2008) calls "projection" wherein the experience of the empathizer and that of the empathizee do not match up. This failure has been observed in numerous contexts when expert-others create meals or diet plans based on nutritional planning or other interests but which the eaters themselves, be they children or food aid recipients refuse. Micah M. Trapp shows how Liberian refugees undermine the biopolitics of humanitarian food aid through refusing particular foodstuffs and through upending the logic of the "taste of necessity" by invoking the necessity of taste through sophisticated culinary transformations of humanitarian food aid as well as through using food to critique their positionality (Trapp 2016). Refugees thus resist the top-down and paternalistic food aid regime that would render them passive recipients. My point here is that nutritional planning has become a common form of biopolitical control across vastly different contexts. Many kinds of eaters refuse the standardization of eating by expressing taste preferences or refusing food altogether.

Through TMR feeding, dairy nutritionists seek to render cows as passive recipients of standardized feed in a form of dietary biopolitical control. Cows undermine the politics of particular nutritional models of diet through expression of particular tastes and through food refusals. Indeed the enormous amount of research on how to make food palatable and the numerous businesses that market solutions certainly seem to upend the logic of cows as passive subjects. Instead cows become the consumers and enormous amounts of effort go into figuring out what they will eat. In addition, they are cultural eaters, what and how they eat is shaped through idiosyncratic preferences but also through Midwestern proclivities towards sweet foods.

Extending Taste

Research in the dairy worlds of Wisconsin reveals how senses, and taste in particular, exceed species-bodies. Taste for cow feed is not reducible either to the tasting bodies of humans or cows but instead is a sense that develops across bodies through a long-term co-presence that enables cow preferences to be understood and that extends cultural associations between care and food provisioning towards cows. As farmers and dairy nutritionists attend to what and how cows eat through observing, tasting, smelling, and touching, they learn transcorporeal taste.

Cows, too, develop taste through learning to eat particular foods and feedstuffs. But they are not passive recipients of feed, despite the biopolitical organization of eating represented by TMR feeding technologies. Cows sort and refuse feed according to personal preferences, social dynamics, and distaste for the flavors and textures of feed components and additives. In this way, taste for cow feed is not at all static, but a continuously evolving transcorporeal form of sensing.

Transcorporeal taste contributes to a growing literature that questions the five-sense model. This model presumes a universal individualized body with distinct and separable sensing capacities. When taste is transcorporeal, it emerges through social interactions not only among humans but among other co-present bodies. It is not a kind of taste that anyone is born with but instead is shaped through encounter. Further, it is not a kind of taste that can be understood through looking only at taste cells in the mouth. Trancorporeal taste extends beyond the mouth and beyond what might seem to be the boundaries of bodies. It brings attention to the locations of taste, the extension of where taste begins, and the co-species training of sensory attention.

Note

1 This material is based upon work supported by the National Science Foundation under Grant No. 1256532; the Danish National Research Foundation and Aarhus University Research on the Anthropocene (AURA); the University of California Santa Cruz (UCSC) Department of Anthropology; and the UCSC Center for Agroecology and Sustainable Food Systems (CASFS). Special thanks also go to Melissa Caldwell, Carole Counihan, Natalie Forssman, Spencer Orey, Jessica Partington, Susanne Højlund Pedersen, Pierre du Plessis, Anna Tsing, and Michael Vine.

References

Caldwell, M.L. (2014). "Digestive Politics in Russia: Feeling the Sensorium beyond the Palate." *Food and Foodways*, 22 (1–2): 112–135.

Callon, M. (1984). "Some Elements of a Sociology of Translation: Domestication of the Scallops and the Fishermen of St Brieuc Bay." *The Sociological Review*, 32 (1_suppl): 196–233.

Classen, C. (1993). *Worlds of Sense: Exploring the Senses in History and Across Cultures*. London: Routledge.

Csordas, T.J. (1993). "Somatic Modes of Attention." *Cultural Anthropology*, 8(2): 135–156.

Desmond, M. (2006). "Becoming a Firefighter." *Ethnography*, 7(4): 387–421.

Despret, V. and B. Buchanan (2016). *What Would Animals Say if We Asked the Right Questions?* University of Minnesota Press.

Grasseni, C. (2007). "Good Looking: Learning to be a Cattle Breeder," in C. Grasseni, ed., *Skilled Visions: Between Apprenticeship and Standards*. 47–66. New York: Berghahn Books.

Grasseni, C.(2009). *Developing Skill, Developing Vision: Practices of Locality at the Foot of the Alps*. New York: Berghahn Books.

Haraway, D.J. (2008). *When Species Meet*. Minneapolis: University of Minnesota Press.

Hayward, E. (2010). "Fingeryeyes: Impressions of Cup Corals." *Cultural Anthropology*, 25(4): 577–599.

Hollan, D. (2008). "Being There: On the Imaginative Aspects of Understanding Others and Being Understood." *Ethos*, 36(4):475–489.

Janeja, M.K. (2010). *Transactions in Taste : The Collaborative Lives of Everyday Bengali Food*. London: Routledge.

Lévi-Strauss, C. (1997). "The Culinary Triangle", in C. Counihan and P. Van Esterik, eds., *Food and Culture: A Reader*, 28–35. London: Routledge.

Mann, A.M., et al. (2011). "Mixing Methods, Tasting Fingers: Notes on an Ethnographic Experiment." *HAU: Journal of Ethnographic Theory*, 1(1): 221–243.

Mauss, M. (2000). *The Gift: The Form and Reason for Exchange in Archaic Societies*. WW Norton & Company.

Mintz, S.W. (1996). *Tasting Food, Tasting Freedom: Excursions into Eating, Culture, and the Past*. Boston: Beacon Press.

Porcher, J. (2011). "The Relationship between Workers and Animals in the Pork Industry: A Shared Suffering." *Journal of Agricultural and Environmental Ethics*, 24 (1): 3–17.

Porcher, J. and T. Schmitt (2012). "Dairy Cows: Workers in the Shadows?" *Society & Animals*, 20(1): 39–60.

Smuts, B. (2001). "Encounters with Animal Minds." *Journal of Consciousness Studies*, 8 (5–6): 293–309.

Thompson, C. (2005). *Making Parents: The Ontological Choreography of Reproductive Technologies*. Cambridge: MIT Press.

Trapp, M.M. (2016). "You-Will-Kill-Me-Beans: Taste and the Politics of Necessity in Humanitarian Aid." *Cultural Anthropology*, 31(3): 412–437.

6

Cultivating a Taste Landscape

Cooperative Winegrowing and Tastemaking in Carema, Italy

Rachel E. Black

"Wine is made in the vineyard!" extolled Stefano as we climbed up through his steep terraced vineyard. As I found my footing at the top of an ancient drystone wall, I began to contemplate what Stefano's statement meant for winegrowers like him. The cooperative Cantina Produttori Nebbiolo di Carema produces nearly all the wine in the small DOC (Denominazione di origine controllata) of Carema[1] and the majority of landowners belong to the cooperative. If wine is really made in the vineyard, how is it that this group of small winegrowers produces the distinctive taste of Carema wine through their viticultural practices? This is the question that I investigated as I spent several months in 2011 clambering up and down the steep hillsides surrounding Carema and interviewing twelve cooperative members and three cooperative employees.

The cooperative, winegrowers, and townspeople pride themselves on the beauty of the cultivated landscape of the vineyards. Pillars, locally known as *pilun*, hold up the tall Nebbiolo vines that grow on the terraced vineyards that are planted up the sides of the mountains surrounding the town of Carema (see Figure 6.1).

The town of Carema and its wine cooperative are closely associated with this landscape—it is an integral part of local identity. Images of vineyards can be found not only on the cooperative's wine labels (see Figure 6.2) but also on posters advertising local events and a variety of materials from the town hall.

The construction of this distinctive landscape is a collaborative project that has taken shape over hundreds of years (Demoissier 2011; Fabian 1983). Although Carema is visually striking, the main sensory goal of this landscape is the production of taste—that of Nebbiolo wine. This small town provides a case study through which to consider the ways in which taste is produced collectively and how a taste tied to a specific locale changes over time yet maintains coherence, both from a visual and gustatory perspective. This chapter shows how the connections between the construction and maintenance of a specific agricultural landscape impact the taste of the wine produced in Carema. Wine has been at the center of explorations of the notion of *terroir*, the taste of place (Trubek 2008), and this case study explores *terroir* as a communal process that

Figure 6.1 Vines growing on typical Caremese *pilun*. Photo by Rachel E. Black. Drystone walls hold up the terraces. The visual effect is stunning in all seasons: walls of lush green in the summer, and the elaborate architecture pergolas revealed in the winter.

Figure 6.2 The label from a bottle of Cantina Produttori Nebbiolo di Carema wine promotes this distinctive agricultural landscape. Photo by Rachel E. Black.

occurs through the production of a landscape, agricultural practices and a local discourse that seeks to bring a dimension of lived experience into the production of taste in a specific locality. This investigation reveals that the negotiation of agricultural practices and the social production of taste are not without their tensions.

While climbing through the vineyards of Carema with Alessandro, another winegrower, he pointed to a neighbor's property, "You see that? My neighbor is breaking with tradition: he is cultivating his vines using the *Guyot* system rather than the pergola. I understand why he does it. He wants to be able to work his vineyard with a tractor. His son moved away and he has no one to help him with the work. I know it is hard but those vines are not the same Nebbiolo that I cultivate. How can we continue making Nebbiolo di Carema when people are changing the way we have always worked?" Alessandro's words began to reveal to me the ways in which a taste landscape is constructed, maintained, adapted, and negotiated.

Historically, farmers who grew grapes and made wine in Carema saw it as both a personal staple and a cash crop that brought in supplemental income in addition to their subsistence farming practices; unlike today, winegrowing was not seen as the only or the most important agricultural product in the area. Traditionally, most families cultivated market gardens, raised a few head of livestock, collected chestnuts and grew cereal crops (generally corn). This tradition began to change during the late-nineteenth century when the steel industry took root near the Dora Baltea River in this area at the base of the Aosta Valley (Borgesio 1980). By the turn of the century, manufacturing began to grow in the area; most notably, in 1908 the Olivetti typewriter factory opened its doors in nearby Ivrea. Local farm labor was drawn to the factories, and the rhythms of life began to change (Argentero 2006). Winegrowing persisted but it became a secondary activity that was pursued between factory shifts and on weekends (Abrardi, et al. 1981). In many ways, industry helped sustain small-scale winegrowing activities since livelihoods were not dependent on wine production—it was more of a hobby, it gave people supplementary income and helped them to maintain their cultural practices such as drinking wine with meals and the local harvest festival (Fornera 1984: 121–122). Employers such as Olivetti gave employees time off during the harvest so that they could tend to their vineyards.

Adriano Olivetti, a strong believer in cooperative agriculture, helped to found the nearby Cantina Sociale della Serra, a cooperative winery, in 1953. Olivetti's *movimento communità*, a political and cultural movement in the Canavese, had a strong impact on Carema and a representative of the "Communità" movement was elected as Carema's mayor in 1956. In 1960, Arturo Perono launched an organization that would protect the production of traditional Carema wine, which was the predecessor to the Cantina dei Produttori Nebbiolo di Carema (Giuliano 1995: 15–16). This grassroots movement aimed to ensure the continuation of the local production of wine, which was particularly important as fewer Caramese cultivated their vineyards because of outmigration and life in factory jobs. In the face of rapidly changing economic realities, the introduction of a wine cooperative in Carema gave a new survival strategy for small producers.

Initially, the winegrowers grew grapes, made wine at home and then transported the finished wine to the cooperative structure, where it was combined, bottled, and sold under the same brand. Viviano Gassino, the president of the cooperative, explained to me during an interview that this first step was necessary because the cooperative did not yet have the capital to build a structure where the growers could make their wine collectively. The Caramesi have a strong tradition of home winemaking and it was central to the social life of the town—the men of the town would spend time visiting each other's cellars to sample wine and socialize, and in some cases important celebrations such as birthdays and confirmations were celebrated with family and friends in the cellar. The cellar was an important productive and social space in the home (Vigliermo 1981). Although located in homes, cellars in Carema have their own entrances separate from the rest of the house. Conversations with older winegrowers in Carema revealed that before the time of the cooperative, Caramese cellars were semi-public spaces where men would gather to drink each other's wine and spend time together. These social moments were for talking about everyday life but also to discuss the qualities of the wine, the weather during the harvest year and any other challenges or successes experienced during the wine-making process. Home cellars were critical spaces for the collective production of taste.

When the cooperative began operating, this separate but collective production of wine posed some serious problems—namely that not everyone's winemaking skills were equal and cellar hygiene was not always a priority. Gassino told me that during the first cooperative vintage after the wine had been bottled, the temperature rose a little in the cellar and the corks began to explode out of the bottles. This meant that not all the winegrowers had allowed their wine to go through mallolactic fermentation. This is a fermentation process that happens after primary fermentation; it turns tart mallic acid in to softer-tasting lactic acid. The exploding bottles were due to this incomplete fermentation process. The cooperative leaders knew that they had to find a way to vinify the wine collectively in order to ensure quality.[2] This marked a major shift in the collective production of taste and social life around wine in Carema.

Gassino told me how in the mid 1960s cooperative members raised the money and contributed sweat equity to build a winemaking facility. Not all members were enthusiastic about the move away from individual winemaking but they understood that in order to continue their membership in the cooperative they needed to contribute all their grapes to the collective facility. These new regulations were introduced in order to ensure the viability of the cooperative and to keep substandard wine from reaching the market and negatively impacting the Carema brand. All members having to contribute the entirety of their grape production also ensured that quality production would be everyone's duty.

The collective production of wine changed the taste of the wine produced in Carema. At first, I had a hard time understanding what my participants were telling me when they told me that the wine no longer tasted the same as when they made it at home. It was challenging to get them to articulate the shift in smell and taste. One man told me that he now had to go to the cooperative to buy his household wine in bulk but it no longer had the vinegar edge that he had come to associate with his own wine. I came to realize that many of my participants had been talking about the move from oxidized homemade wine to the cleaner taste of wine that had been made and preserved in more hygienic conditions. When wine kept in large tanks is exposed to too much oxygen, the anthocyanins and phenols begin to oxidize, which leads to a loss of color and a flattening of the taste. Nutty, sherry-like notes replace the aromas and flavors of the original fruit and tannin of the wine. Bottling wine, which limits exposure to oxygen, helps put a stop to this process. In addition, a lack of cellar hygiene can lead to growth of spoilage yeasts such as Brettanomyces. This yeast can give a barnyard aroma to wines and can potentially lead to even less desirable aromas such as Band-Aids.

The technological changes in wine production were hard to swallow at first. As I discussed this point with a variety of participants, they explained that it was just something that they got used to over time. This challenged my idea that taste was something that could have coherence over time. Although the local Nebbiolo grapes had typical characteristics of high acidity and tannin, the overall taste profile had changed over time, largely due to technological developments in production methods. In this case, taste memory is malleable (Black 2012). This case shows that there are moments of resistance that are followed by a slow acceptance and changing collective articulation of the wine's taste. The Caramesi winegrowers moved from having individualized taste memories of the local wine to developing a collective understanding of the taste of the local wine as the vinification process moved out of their cellars. Individuals were no longer in charge of the winemaking process and therefore lost their ability to make their mark on their own unique wines. At the cooperative's new winemaking facility, a professional winemaker was responsible for crafting the taste of Carema wine.

Although cooperative members offered their knowledge and expertise, a new regime of scientifically driven winemaking came into play: Carema wine began to respond to the demands of the broader Italian and international quality wine markets. These markets demanded that wines be free of faults related to poor cellar hygiene and practices. In this case, the outside market, which Carema needed to be a part of in order to financially survive, began to dictate norms and expectations of taste associated with the Nebbiolo grape variety. At the same time, the cooperative winemaker had to express the unique *terroir* of Carema. This process was a sort of forced unification of taste that was born out of a need to collectively produce wine to reach economies of scale and a need to conform to practices and standards of quality that the larger wine market required.

Cooperative winemaking was a big shift from the showcasing of individual identities through winemaking skill to a collective pride of place through the cooperative production of one wine to represent the labor of all the people in one place. This was not a seamless process that erased all traces of individual labor and skill. As I mentioned earlier, economic duress made this shift from the individual to the collective a necessity. At the same time, the cooperative and town hall worked to move the emphasis from individual winemaking to individual grape growing. The Festa dell'Uva, which is held each September, is now in its sixty-sixth year. During this five-day celebration, there are a number of events that highlight local culture—art exhibits, craft markets, and concerts. The pinnacle of the celebration is the Saturday night tour of historic *cantine*. The Caramesi open their homes to visitors to see their cellars that they immaculately keep as unofficial museums to the town's winemaking past. Wine and food are central to the tour, but all the wine is now made at the cooperative. On Sunday, the festivities conclude with a public ceremony for the awarding of the *Grappolo d'Oro* (the golden grape bunch). This award represents the shift from the celebration of individual wine production to grape growing. A panel of judges give the *Grappolo d'Oro* to the grape grower who presents the most beautiful bunch of grapes. The focus is firmly placed on the agricultural side of wine production. The grape grower is celebrated as the source of quality in the production of wine in Carema.

The quality control of grape production is a social process in Carema that symbolically takes place during the Festa dell'Uva, but on a more practical level at the cooperative winery. Fabio, a cooperative member, explained to me that when members bring in their grapes, they are first processed on the sorting table. There is always a group of members in the sorting an area. They hang out, socializing unofficially keeping an eye on the sorting table. When a member brings in nice grapes, the group praises the grower. I was told that no one dares bring in low-quality grapes—all members fear the wrath of the communal gaze at this critical moment in the winemaking process (Wolf, et al. 2001). It is not unheard of for a member who brings in damaged grapes to be publicly shamed. This is where the social element of cooperative production ensures the quality of the base product. When I asked one winegrower about bringing in grapes to the winery, he told me, "It is the moment of truth. You can't hide rotten grapes on the table. If I have some grapes that are not very nice, I leave them in the vineyard to feed the birds." The sorting table is the moment of judgment after all the work and trials in the vineyard throughout the growing season. The sorting table is also another important place where cooperative members collectively negotiate quality and taste—delicious wine cannot be made without excellent grapes.

By collectivizing the winemaking process the cooperative was slowly able to realize economies of scale, enabling it to compete with larger producers while still maintaining local agricultural traditions. Many of the participants in this research reasoned that without the cooperative both grape growing and winemaking would have died out in the area long ago. The Cantina dei Produttori Nebbiolo di Carema has operated with moderate success for much of its fifty-eight-year history, but in recent decades it has begun to face new challenges. After the failure of the heavy industries that had bolstered the local economy, the Canavese went into a period of general economic decline in the 1990s. Out-migration has left an aging population to tend the area's historic vineyards (Popolazione Carema 2001–2015). Due to this lack of renewal in the labor force, the cooperative is now taking a prominent role in the preservation of the landscape, the vineyards, and the cultural practices associated with viticulture—from the collective working of abandoned vineyards to the maintenance of historic viticultural practices. Wine cooperatives in this area of Northern Italy have shifted from performing a strictly economic function (structures that make and market wine) to having a central role in the preservation of the living cultural heritage associated with wine production and the maintenance of the environmental integrity of the terraced slopes. Since the 1960s the cooperative has worked to mediate and support the practices of the winegrowers but in recent years these interventions have become more involved. During a recent visit to the cooperative, the administrator told me that they now had an employee who worked the vineyards of members who were unable to do so themselves. They hoped that this would help maintain the production area in the face of their aging membership. In many ways the cooperative increasing its involvement has created greater coherence in taste through higher standards for the quality of fruit and it has also introduced new viticultural techniques and technologies into the historic vineyards of Carema. Like other winegrowing regions, Carema has had to balance the tensions that exist between tradition, innovation, and the gustatory demands of the international wine market.

One way in which wine regions mediate these tensions is through the creation of regulations that protect traditional practices in production and attempt to conserve the typical organoleptic characteristics of an area's wines. In 1967, the Cantina Produttori Nebbiolo di Carema undertook the creation of the Carema *Denominazione di Origine Controllata* (DOC).[3] The cooperative executive played a key role in determining the geographic area of the DOC and the strict production practices for the DOC wine. Large economic interests often drive the creation of these protected statuses for food and wine (Gade 2004). In the case of Carema, there were economic motivations but there was also a desire to ensure quality and to protect the production area. These objectives validate the cooperatives' work to produce and promote the sale of quality wines. Maby suggests that with an increasing focus on the production of quality wines, winegrowers have begun to consider the production of the aesthetic of the winegrowing landscape as a value-producing element of their work. Where once the creation of a specific landscape was a spontaneous and reflexive part of producing wine, the esthetics of place have now become conscious and intentional on the part of winegrowers as they come to consider the image of the place and its importance for marketing their product and for attracting tourists (Maby 2002). With the emphasis on the notion of *terroir*, this is particularly true for quality wines that distinguish themselves based on unique microclimate and geological specificity (Bérard and Marchenay 1995).

The Carema DOC stipulates that traditional clones of Nebbiolo must be used in the production area but it does not specify precisely which clones. A clone is a cutting taken from a "mother" vine which is then used to propagate a new plant. This new plant will have the same genetic makeup as the "mother" vine. In winegrowing, clones are selected for disease resistance, quality of grapes produced, their suitability to a specific site, among other considerations. Over time, winegrowers generally come to prefer specific clones that have done well in the area where they are cultivated. Different clones of the same grape variety can vary widely in taste (Robinson and Harding 1994: 237). In addition, the regulations were changed from their original iteration to permit Guyot cultivation ("sistema di allevamento a spalliera") for new plantings—a big departure from the traditional pergola method that generally requires clones that are adapted to this training system (see Figure 6.3). The nineteenth-century viticulturalist Jules Guyot developed a method of cane pruning where each vine has one or two canes preserved each year from which the next generation of next year's fruiting canes will come, and one spur, where the replacement cane will come from. The canes are trained flat against a trellis system. In contrast to the Guyot method, the pergola requires head-trained vines that have tall trunks with a spur-pruned system with vines reaching up into an overhead trellis (Robinson and Harding 1994). This is an ancient method of vine production that was favored by the Romans, some of the early producers of wine in the Carema area.

The reasons for this change are practical: Guyot training allows for the mechanization of much of the cultivation process, it facilitates pruning and harvesting. With the Guyot system, the winegrower can drive a tractor through the vineyard. This expedites the

Figure 6.3 Nebbiolo grapes grown using the Guyot training method. Photo by Rachel E. Black.

application of treatments against pests and the harvesting of grapes. Although the Carema terrain is steep, a reorganization of vines, using the more modern Guyot system would allow for cost-saving mechanization. In the end, an argument could be made for the shift in farming methods based on economics and quality. However, there are also strong arguments about aesthetics that support staying with the old pergola system. First, new clones of Nebbiolo suited for growing on the Guyot system need to be selected. The current clone is the product of the selective breeding of vines over a long period of time, and the introduction of a new clone would be a more rapid evolution of this adaptive process. Some Caremesi winegrowers point to a loss of genetic heritage through this shift. Second, visually these two systems are extremely different. If there is a sea change towards the Guyot system, the aesthetic landscape will no longer be the one the cooperative's labels or the one used by the town to promote cultural and tourism events. The group of Guyot upstarts, a minority among cooperative members, argue that this is the only way that they can viably continue to grow grapes.

It is apparent that through this allowance of different training methods the Carema DOC does little to protect the tangible cultural heritage of the vineyard landscape; rather, it focuses on oenological practices that conform to international standards for the production of quality wine. This includes ensuring a certain length of aging before release (twenty-four months for the "Carema" DOC wine and thirty-six months for the "Carema" reserve) and minimum alcohol levels (Regione di Piemonte). This lack of regulation for cultivation methods is challenging the coherence of local agricultural methods, and in turn the visual aesthetic of the landscape. Levin argues that the move from pergola to Guyot vine training systems aids vineyard management and helps improve the quality of the grapes produced, but he is unable to make a strong connection between the actual improvement of a wine's taste and these agricultural changes (Levin 2015). It is still early days and it remains to be seen what impact these changes will have on the taste of the wine as more and more grape growers move from pergola systems to Guyot.

Although the Carema DOC protects a specific historic area of grape production, it has done very little to help the cooperative maintain the number of hectares in production and the unique aesthetic of the landscape. Each year the amount of cultivated terrain shrinks a little as winegrowers abandon their vineyards. Many are too old to work, or find the work too arduous and others have left the region to find jobs. This has been a major challenge for the cooperative, which has 101 members who cultivate a total production area of 13.5 hectares. The records show that at the beginning of the twentieth century there were as many as 120 hectares under cultivation (Giuliano 1995). The main decline occurred after World War II and only stabilized in the last five years. As demand for Carema wines has increased, the cooperative struggles to meet the market demand. As mentioned earlier, the cooperative has begun to hire workers to work members' plots as a strategy for maintaining production. This initiative has started slowly in the past four years. Initially, the cooperative did not think it would be profitable to pay for employees to work the land and do the harvest. In fact, if the cost of production is calculated at fair market prices, Carema wines would be too expensive to produce given the current retail price. Carema is an economically impossible wine. Today's production counts on the unpaid or uncalculated labor of families, not an uncommon practice in Italian agriculture (Calus and Van Huylenbroeck 2010). This is clearly an unsustainable situation from an economic perspective, and it is also an issue that has raised questions among cooperative members. Will the paid labor carry on the traditional practices of caring for the vineyards? How will this impact the taste and quality of the wine? Most of the workers are migrants and cooperative members were left wondering how these newcomers would integrate into the community? What changes would they bring to grape growing practices?

It is difficult to say whether or not the cultivation method and the introduction of new clones will impact the final taste of the wine. However, for those who believe that wine is made in the vineyard, these new developments are contentious and threatening. The cooperative is going through a period of dissension among its ranks as members test different strategies for maintaining production. One member told me, "It is better to go forward and to keep producing grapes, rather than giving up and letting the fields grow over. Sometimes this means making changes, even if you don't like them." Some changes in production seemed easier to accept than others for cooperatives members.

One change I was surprised to witness was that of the use of small oak barrels for aging Carema wine. During my fieldwork in 2011, the cooperative winery was moving to aging a number of its wines in *barrique* (225–228 litres), which was a big move away from the traditional *botti* (made of oak, walnut and maple and holding thousands of litres of wine) that had been used for hundreds of years. The cantina cooperativa was only moving a portion of its production to the smaller barrels in an attempt to appeal to New World markets where wine is consumed young and consumers have a preference for riper fruit and softer tannins. Nebbiolo wines have long had an issue finding favor in the American market because of their fierce tannins when young. The use of *barriques* to soften young wines has been contentious for other renowned Nebbiolo producing areas, such as Barbaresco and Barolo (Hannan, et al. 2007).

This change in production method is a clear break from tradition and it has a big impact on the flavor profile of the wine. The use of *barriques* in Piemontese winemaking is deeply influenced by trends in the international market. It was surprising to me that Carema was trying out this experiment in 2011 when so many other Nebbiolo producers had tried using *barriques* and had gone back to large *botti* after not being satisfied with the aging capacity of their wines. *Barriques* had seemed like a fleeting fad (Corrado and Odorici 2009). Trying to better understand this experiment, I spoke at length with Manlio, the cellar master at the cooperative, and he told me that the cantina's consulting winemaker had encouraged this change as a way to tap into new markets. The proposition came at a time when the sales of Carema wines were declining and cooperative members were open to new strategies that would help them sell more wine. I was surprised that members would agree so readily to such a dramatic shift away from traditional practices, but I also took into account that this was only a change to a small portion of the production. It also occurred to me that perhaps members were more open to changes in production methods in the cellar versus the vineyard because they had been largely removed from the vinification process forty years before. As far as member agreement on what Carema wine should taste and look like, these new wines were not really part of their vernacular because these were not wines that they were drinking together. The new wines were marketed under a separate label and had a markedly higher price. When I asked one member what he thought of the new wine, he told me, "Oh, that is wine for Americans. We don't drink that here."

During many interviews and informal discussions, cooperative members told me about earlier changes in taste that were slowly accepted when the cooperative moved to a collective vinification process. The Caremesi had to get used to their wines not having a slight vinegary taste and less microbial variation. At first, they did not recognize their own wine but over time they adapted. As more grapes grown on Guyot are introduced into the mix, will there be a similar gustatory shift? As the first instance shows, the collective taste memory of wine is mutable. Some participants in my study talked about sensory *fili conduttori* (guiding lines) that remained strong despite the technological shifts in production (Black 2012). For example, some winegrowers described the wine as having the same characters as the mountain people who grow it, austere and rigid, and that you could taste the place through other plants that grow among the vines, such as *mentuccia* (*Calamintha nepeta*), in the wine. In these descriptions, taste is linked to the character of people in the place and to the natural environment. It is difficult to say at this point whether these taste frames of reference will shift.

I came to understand that this way of talking about the taste of wine in reference to place is a social process (Parkhurst Ferguson 2011). The language of taste is negotiated and codified in the private cellars under people's houses. Although these historic places are no longer used for wine production, they are an important place where people, mostly men, spend time together socializing and drinking. This is where people talk about life, Carema, and wine. In these *cantine* important life events, such as coming of age celebrations, marriages, baptisms, are celebrated with members of the village. The language of wine is not that of the formal wine tasting world that focuses on the wine trade. It is a language that talks about place, relationships to the land, and a collective belonging to a community.

This is what I experienced as I got to know several Caramese winegrowers. Giuseppe gave my husband and me a thorough tour of his vineyards: up and down the steep slope with stops along the way to point out ancient property markers, a few negative remarks about neighboring plots using the Guyot system, sad comments about nearby vineyards that had been abandoned, and a pause to admire a wispy willow that served to provide organic ties for training vines to pergolas. We came to what we thought was the end our visit in the village near Giuseppe's house. However, we were not done yet—our guide asked us if we would like to drink some wine. I noted too that Giuseppe had not asked us to come taste (*assaggiare*) his wine. He had used the verb to drink (*bere*). We happily accepted Giuseppe's invitation and proceeded to a side door of his house and down a flight of stairs to a beautiful cantina with woodcarvings and ornamental barrelheads. Giuseppe dug around in his stash of bottles in an area behind the main room. He emerged with a dusty bottle of 1968 Carema. He quickly told us that he could make no promises about the bottle and then he began to go to work opening it. The fragile cork began to crumble and we all held our breath hoping that the liquid in the bottle would still be wine and not vinegar. Giuseppe wiped the rim of the bottle and poured out the rust-colored juice into our glasses. Giuseppe quickly sniffed and drank. A smile came over his face. We hesitantly put our noses to our glasses and slowly inhaled. Immediately, we were struck by the vitality of the wine. We began to smile, anticipating the delights held in our glasses. We finally sipped and then sat back in awe of the longevity of one of Italy's great wines. Giuseppe puffed out his chest with pride, and told us the story of this vintage and how he had worked with his father making the wine. We continued to happily sip our wine and chat about winegrowing, until we had finished the bottle. Very few words were used to talk about the taste of the wine. The narrative of taste focused more on a shared understanding of the work that had gone into producing this bottle of wine. We savored the last drops, feeling a connection with Giuseppe. He had just shared with us a precious and rare piece of his cultural heritage.

That afternoon, I came to understand that drinking wine together is very different from tasting wine together. At the cooperative tasting room, I had tasted the most recent vintages of Carema wines. Salena, the office manager, gave me small pours in a tasting glass. I went through the usual motions of evaluating the color, smelling the wine and noting the aromas and then finally tasting the wine, drawing air in to my mouth in slurps to make sure all of my mouth sensed the wine. I then spit the wine into a spittoon. This is very standard practice in the winetasting world. This professional practice of winetasting is an attempt to be objective and rigorous about evaluating subjective knowledge in reference to previous experiences (Peynaud 1995). In addition, the language used to talk about wine in a professional setting like a tasting room draws on a rich and poetic sensory vocabulary that is guided by an increasingly standardized system of wine aroma terminology (Noble, et al. 1984). A tasting note from *Decanter Magazine* for the 2007 Cantina dei Produttori Nebbiolo di Carema, Classico reads as follows:

> Delicate but complex nose of red cherry, mint, aniseed and tea. Dry, intense palate with fine tannins and some oak still to integrate. Firm, grippy, mineral finish, bitter herbs and orange peel.
>
> Hindle 2012

This tasting note demonstrates the ways in which the sensory experience of wine takes this beverage out of the social context that gives it meaning in order to assign a new and seemingly objective value related purely to aesthetics. The tasting note is a necessary communication tool when trying to evaluate and sell wine, but it is an abstraction of a richer narrative that belongs to the people and the place where the wine is produced—this is the story of the taste landscape and the language that is collectively cultivated along with the vines.

In the context of sharing wine with growers, a professional approach to tasting would have been offensive. How could someone spit out the hard labor of an entire family? Was wine not for sharing? "Wine is for being happy together," one grower explained after another cantina session. The production of taste also takes place in this context of drinking wine together. Rather than the objective appraisal of wine, taste is articulated subjectively through its function of creating social moments. The wine is talked about and evaluated but in terms of labor and memories of the growing season of the specific vintage. These are the social elements that are critical in negotiating *terroir*. The ties between production practices, climactic factors, personal situations and the final product of the wine in the bottle make more sense in this familiar, social context. The way in which winegrowers evaluate taste moves beyond the attempts of professional wine tasters to create an objective definition of taste—they were producing a socially embedded and geographically specific notion of taste. These are elements that are not often considered in more conventional definitions of *terroir* (Wilson 1998).

Tradition guides the cultivation of Nebbiolo grapes in Carema. However, traditions are constantly challenged and viticultural practices are negotiated between generations and emerging schools of thought (Hobsbawm and Ranger 1983). Carema offers a case in point for understanding the instability of traditions. The question remains about how long it will take the community to adapt to and adopt new technologies and ways of doing. When does Guyot become tradition, if it does at all? Will a new terraced landscape be featured on the cooperative's wine labels or will the identification through a unique landscape become secondary for the Caremesi?

At present, the need to innovate and change cultivation methods in Carema is particularly pressing as the aging population of winegrowers and the cooperative grapple with the lack of labor and a declining local population. The case of Carema shows how the cooperative construction of taste is not always harmonious. Nonetheless, the lived realities I was able to witness in this small subalpine village and the stories I heard from the people there showed me how the common cultivation of land and the narratives surrounding this labor can lead to shared notions of taste. The concept of *terroir* benefits from a closer look at the negotiation of the social relations that mediate the production of taste.

Notes

1 Disciplinare di produzione dei vini a denominazione di origine controllata "Carema", Retrieved from: www.regione.piemonte.it/agri/politiche_agricole/viticoltura/dwd/disciplinari/carema.pdf
2 In this case quality means a certain industry standard that carries the expectation of full fermentation and none of the negative associations that secondary fermentation creates, such as effervescence or commercial losses due to exploding corks.
3 The DOC is similar to the French AOC (Appelation d'Origine Contrôlée). It is a quality assurance label introduced by the Italian government in 1963. It designates the area of production and specific production methods that must be used in order for a wine to be certified with the DOC label. In 1993, Italian DOC laws were brought into conformity with EU regulations on protected geographical designations of origin. http://eur-lex.europa.eu/LexUriServ/LexUriServ.do?uri=CONSLEG:2002R0753:20071214:EN:PDF

References

Abrardi, G., L. Vittori and A. Vigliermo (1981). Carema, gente e vino. Ivrea: Priuli and Verlucca.

Argentero, R. (2006). Quando in Canavese esistevano le "grandi" fabbriche. Ivrea: Hever Edizioni.

Bérard, L. and P. Marchenay. (1995). "Lieux, temps et preuves." *Terrain*, (24), 153–164.

Black, R. (2012). "Taste Memory." *Sensate Journal*. Accessed January 11, 2017, from http://sensatejournal.com/2012/06/rachel-black-wine-memory/

Borgesio, M. (1980). Carema, origini storiche e sviluppo economico. Montalto Dora: tipografia Gianotti.

Calus, M. and G. Van Huylenbroeck. (2010). "The Persistence of Family Farming: A Review of Explanatory Socio-economic and Historical Factors." *Source Journal of Comparative Family Studies*, 41(5), 639–660.

Corrado, R. and V. Odorici. (2009). "Winemakers and Wineries in the Evolution of the Italian Wine Industry: 1997–2006." *Journal of Wine Research*, 20(2), 111–124.

Demossier, M. (2011). "Beyond Terroir: Territorial Construction, Hegemonic Discourses, and French Wine Culture." *Journal of the Royal Anthropological Society* 17: 685–705.

Fabian, J. (1983). *Time and the Other: How Anthropology Makes its Object*. New York: Columbia University Press.

Fornera, I. (1984). "Carema: di tutto un po." Ivrea: Tipografia-Litografia Bolognino Davide & C., s.n.c.

Gade, D.W. (2004). "Tradition, Territory, and Terroir in French Viniculture: Cassis, France, and Appellation controlée." *Annals of the Association of American Geographers*, 94(4), 848–867.

Giuliano, W. (1995). Ricerca sull'evoluzione a memoria d'uomo, della tecnica e del linguaggio viticolo-enologico in centri rappresentativi del Piemonte: no. 7 Carema. Torino: Associazione Museo dell'Agricoltura del Piemonte.

Hannan, Michael, Giacomo Negro, Hayagreeva Rao, and Ming De Leung. (2007). "No Barrique, No Berlusconi: collective identity, contention, and authenticity in the making of Barolo and Barbaresco wines," Stanford University Graduate School of Business Research Paper No. 1972. Retrieved January 12, 2017, from https://papers.ssrn.com/sol3/papers2.cfm?abstract_id=1008367

Hindle, G. (2012). "Uncharted Italy: Carema DOC, Piedmont," *Decanter* Magazine. Retrieved May 29, 2017, from www.decanter.com/wine/wine-reports/unchartered-italy-carema-doc-piedmont-31731/

Hobsbawm, E.J. and T.O. Ranger, eds., (1983). *The Invention of Tradition*. Cambridge: Cambride University Press.

Levin, D. (2015). "When do wines taste different. Part 2: Vineyard management," Cool Climate Wine. Retrieved May 29, 2017, from https://coolclimatewine.wordpress.com/2015/08/08/why-do-wines-taste-different-vineyard-management-part-2d/.

Maby, J. (2002). "Paysage et imaginaire: l'exploitation de nouvelles valeurs ajoutées dans les terroirs viticoles." *Annales de Géographie*, 111(624), 198–211.

Minardi, M., and E. Franchetto, (1960). Il Canavese ieri e oggi. ILTE.

Ministero delle Politiche Agricole e Forestali. (2010). "Disciplinare di produzione dei vini a Denominazione di origine Controllata 'Carema.'" (G.U. n° 234 del 6 ottobre 2010). Retrieved January 11, 2017, from www.regione.piemonte.it/agri/area_tecnico_scientifica/osserv_vitivin/dwd/disciplinari/carema.pdf

Noble, A., R.A. Arnold, B.M. Masuda, S.D. Pecore, J.O. Schmidt and P.M. Stern. (1984). "Progress Towards a Standardized System of Wine Aroma Terminology." *American Journal of Enology and Viticulture*. January 1984 (35): 107–109.

Parkhurst Ferguson, P. (2011). "The Senses of Taste." *The American Historical Review*, 116(2), 371–384.

Peynaud, E. (1995). *The Taste of Wine: The Arts and Science of Wine Appreciation*. Trans. M. Schuster. New York: J. Wiley & Sons, Inc.

Popolazione Carema (2001–2015) Grafici su dati ISTAT. (n.d.). *Tuttitalia.it*. Retrieved January 12, 2017, from www.tuttitalia.it/piemonte/21-carema/statistiche/popolazione-andamento-demografico/

Regione di Piemonte. (2011). "Disciplinare de produzione dei vini a denominazione di origine 'Carema.'" Retrieved May 25, 2017, from www.regione.piemonte.it/agri/politiche_agricole/viticoltura/dwd/disciplinari/carema.pdf

Robinson, J and J. Harding, eds., (1994). *The Oxford Companion to Wine*. Oxford: Oxford University Press.

Trubek, A. (2008). *A Taste of Place: A Cultural Journey into Terroir*. Berkeley, CA: University of California Press.

Vigliermo, A. (1981). *Carema: gente e vino*. Ivrea: Priuli & Verlucca, editori.

Wilson, J. (1998). *Terroir: The Role of Geology, Climate and Culture in the Making of French Wines*. Berkeley, CA: The University of California Press.

Wolf, S., B. Hueth and E. Ligon. (2001). "Policing Mechanisms in Agricultural Contracts," *Rural Sociology*, 66(3), 359–381.

7

Tasting Comté Cheese, Returning to the Whole

The *Jury Terroir* as Ritual Practice

Christy Shields-Argelès

It's a sunny February afternoon in the small town of Poligny, France, and the jury terroir *is about to begin. Eighteen people are seated around a set of rectangular tables arranged in a U-shape. "Well there are a lot of people today!," Florence exclaims as she enters the tasting room. The buzz of small talk subsides as participants look around at one another, smile and nod in agreement. "Let's welcome back our new members," Florence says as she acknowledges two farmers who have recently joined the group. "And we have two visitors today," she continues, "Everyone knows Jean's wife I believe? She is judging a cheese competition next week, and so brushing up on her skills. And then Christy is here, anthropologist, as you know".*

This small group of volunteer tasters has met approximately ten times a year, for the past twenty-five years, to describe the tastes of Comté cheese. They are a diverse group of women and men, from the Comté chain and the region, ranging in age from their early twenties to their late seventies. Some have been members since the group's inception in the early 1990s, while others have only recently joined. As an anthropologist, I aim to translate, into relevant and thought-provoking terms, the life ways and worldviews of others. I am currently exploring the taste practices of the Comté chain from an ethnographic perspective because I feel that they address important questions about *terroir*, taste, and the struggle of artisanal food producers to endure and thrive. In 2013–2014, I attended six *jury terroir* sessions as a participant-observer and carried out twenty-two in-depth interviews with past and current jury members. In this chapter, I approach the *jury terroir* as a ritual practice, and argue that it sustains a shared belief in *terroir* through the structured sharing of aromatic descriptors.

Comté Cheese: A *Terroir* Product, A Diversity of Tastes

Comté is a cooked and pressed raw-milk cheese produced in the Jura Mountains in eastern France. While much has been written about Comté's exemplary status as a *terroir* product, what is often lost in those texts is the fact that no two wheels of Comté are alike. This sensorial diversity is due to the manner in which Comté is produced;

there are, for example, a set of strict production rules that maintain the link between place and product, and the cheeses are produced in a web of village-based cheese dairies, or *fruitières,* which are cooperatively owned and operated by the farmers. Bowen (2011) identifies the Comté chain actors' shared belief in *terroir* as an important form of embeddedness that helps them to resist and adapt to the disembedding processes associated with globalization. Bowen notes that this belief can be expressed in a number of ways, and my own research confirms this: *terroir* is associated with the richness of the milk, the *fruitière*'s dairy basins (the soils, microclimates and so on that surround each *fruitière*), and the region generally. Above all else, however, Comté chain actors talk about *terroir* in terms of diversity.

Bowen attributes this shared belief in *terroir*, and adherence to the value of diversity, to French political institutions and culture as well as to initiatives like the *Terroir Program*, a scientific program initiated by the CIGC[1] and directed by food scientist Florence Bérodier. Within the frame of this program, Bérodier and other scientists drew together sensorial characterizations of the cheese, soil maps, assessments of climatic conditions and inventories of plant species in order to produce a number of publications that successfully link the tastes of Comté cheese to the places of their production, thus providing scientific data in support of *terroir* (e.g. Monnet et al. 2000). Such findings are also presented in the chain's bi-monthly newsletter (*Les Nouvelles du Comté)* in the form of a *fruitière* "identity card". In each issue, one specific *fruitière* is featured in a two-page spread that integrates scientific data alongside other sorts of information about the *fruitière*, such as descriptions and pictures of its farmers. The *jury terroir* is integral to all these publications since it is literally the participants' senses that are mobilized in the construction of the sensorial characterizations and aromatic profiles that appear in them.

However, while the *jury terroir* plays an important role in mapping out the "tastes of place" of Comté, it must also be recognized as a taste language and practice that originally helped to establish, and now helps to maintain, a shared belief in *terroir*. While Comté cheeses have always had diverse tastes, *terroir* was not a term used by chain actors to speak about diversity prior to the 1990s. Original jury members were skeptical about the notion of *terroir* (and its applicability to a cheese), as well as this kind of taste practice, which was suspected of being either elitist or nonsense. In its early years then, the jury *terroir* was a key site for the development of a belief in *terroir* (Shields-Argelès 2016). In this chapter, I would like to consider how their taste language and practice promote an experience of the whole that continues to reflect and sustain this same belief.

Terroir and Taste

Social scientists have contributed to a recent interest in *terroir* by arguing that it should be understood as a worldview that links the natural and cultural worlds. To produce and consume *terroir* products, is to think and act holistically, with an attention to interrelation. While many scholars of French *terroir* have investigated interrelation at the discursive and systemic levels of analysis—examining, for example, its historical development (e.g. Trubek 2008), or institutional and regulatory structures (e.g. Barham 2003)—another smaller set of scholars has examined, instead, the daily practices and interactions through which *terroir* is experienced and embodied.[2] Leynse (2006), for example, considers how French children become "situated eaters" as they travel through "ingestible topographies" on fieldtrips. Working in another cultural context, Paxson (2012) suggests that tasting, based on a shared vocabulary and interactive exchange, helps producer and consumer express the moral value of craft production. I hope to contribute to this literature by focusing on tasting as a place-making practice, where "place" (like the notion of *terroir*) is approached as the dynamic, and sensual, relationship between persons and their environment (Feld and Basso 1996).

To develop this point of view, I take initial inspiration from the work of Hennion and Teil (2004) who approach taste as a reflexive, framed and co-constructive activity of the "amateur"; an actor they locate between professionals, with their "certified expertise," and laymen, with "their lack of attention to what they eat" (2004: 19). The amateur does not attend to evaluation but consciously builds attachments to a repertoire of loved objects, and during this process evolves along with the object itself. This theoretical frame fits, first, because the *jury terroir* was designed to attend to description, as opposed to other tasting practices in the chain that focus on evaluation (e.g. grading cheeses for quality control). Likewise, both Bérodier and jury members identify as amateurs; in the words of a farmer: "the team, the people who come here, are all amateurs, Comté lovers."

However, because we are interested in exploring the manner in which this co-constructive tasting activity embeds both taster and tasted in place, we require a theoretical frame that reaches beyond a focus on amateur practice alone, in order to explore the relationship between tasting practices and wider contexts. In this chapter, drawing from the works of Fernandez and Sutton, I approach the *jury terroir* as a ritual practice that helps individuals to recall and relate to the wider social and natural world through the repeated sharing of aromatic descriptors.

A Reflexive Anthropology of the Senses

Before moving on to the analysis proper, the reader should understand that the interpretation elaborated in this chapter is also informed by a particular set of life experiences and my liminal subject position. While born and raised in the United States, I have lived and worked in France for twenty years. My relationship with the Jura region goes back to my first stay as a study abroad student in 1990–1991, and for the past decade I have organized study trips to the land of Comté for my students at the American University of Paris. I have therefore participated in a diverse array of tastings in the Jura region, and elsewhere in France. I am not a particularly gifted taster, but I have long enjoyed tasting for the feelings of connection it affords with others, be they people, foods or places. However, over the years I have also been privy to the various interpretations made of tasting events by visiting Anglophones (especially students) who as visitors are unable to, or uninterested in, establishing and maintaining social ties through tasting. Instead, they often tend towards an experience of the French

Other that is tacitly bound to their own identity construction. Within this dynamic, tastings are experienced as the incorporation of a French *art de vivre*, an encounter with French elitism, a blast to an idealized past, or simply too different (e.g. too raw, too gooey, too smelly) to engage. All of which, to be honest, are positions that I have also occupied at one time or another.

Renato Rosaldo (1993) recounts how acquiring a new subject position allowed him a shared emotional experience, or "co-feeling" (174), with his informants, encouraging him to turn away from the analysis of rituals as fixed routines and approach them instead as open-ended, emotional processes. Similarly, my subject position, and the manner it has shifted over the years, encourages me to examine the tasting practice of the *jury terroir*, not as a set of socially prescribed behaviors in which individuals have little choice, but as a framed and shared process characterized by an emotional experience of connection and transcendence.

Of course, I fear (as Rosaldo did) that this approach, and its tie to my personal experience, may be interpreted as too subjective or naïve. Yet, I evoke my personal experience here for two reasons. First, I subscribe to a reflexive approach to anthropology because without it fieldwork can too easily remain trapped in historically and culturally shaped perspectives that remain unacknowledged. The transformative potential of tasting has perhaps not been given adequate attention in the literature due, in part, to the fact that tasting, especially when practiced in France, is all too readily associated with a rather narrow set of internal cultural processes (e.g. the celebration of French national identity, the reproduction of class hierarchies). Relatedly, the reader should understand that the experiences and feelings revealed here were produced within intimate and dialogical conversations between amateurs who share an experience of and a love for tasting, as well as for Comté cheese. In short, from my particular subject position, I had access to and an interest in experiences and emotions that jury members rarely share with others, and yet are central to what makes the practice deeply meaningful for them. It is my suspicion that such experiences and emotions are familiar to amateurs tasting in other cultural contexts as well.

"The words of the *terroir*": Aromatic Descriptors as Symbols

The jury's aromatic descriptors are used to describe a cheese's aromas, or the smell of a cheese once it is in the mouth. The aromas are one aspect of a cheese's overall taste. The aromatic descriptors, chosen and appropriated by the *jury terroir* in its early years, also appear on the Comté Aroma Wheel (Bérodier et al. 1997).[3] In an earlier publication (2016), I note the descriptive nature of this language. Here, I draw the reader's attention to the figurative nature of this same vocabulary (see Figure 7.1).

To describe the smell of a cheese one must use metaphors. Sutton (2001) demonstrates that smells are "symbols par excellence" (90), first, because they "evoke what surrounds them in memory, what has been metonymically associated with the smell in question" (89); and, second, because smells are strongly associated with "episodic" memories and so adept at recalling emotionally-charged wholes. Smell memories can be individual (as in, hay reminds Jane of her grandfather's farm), but

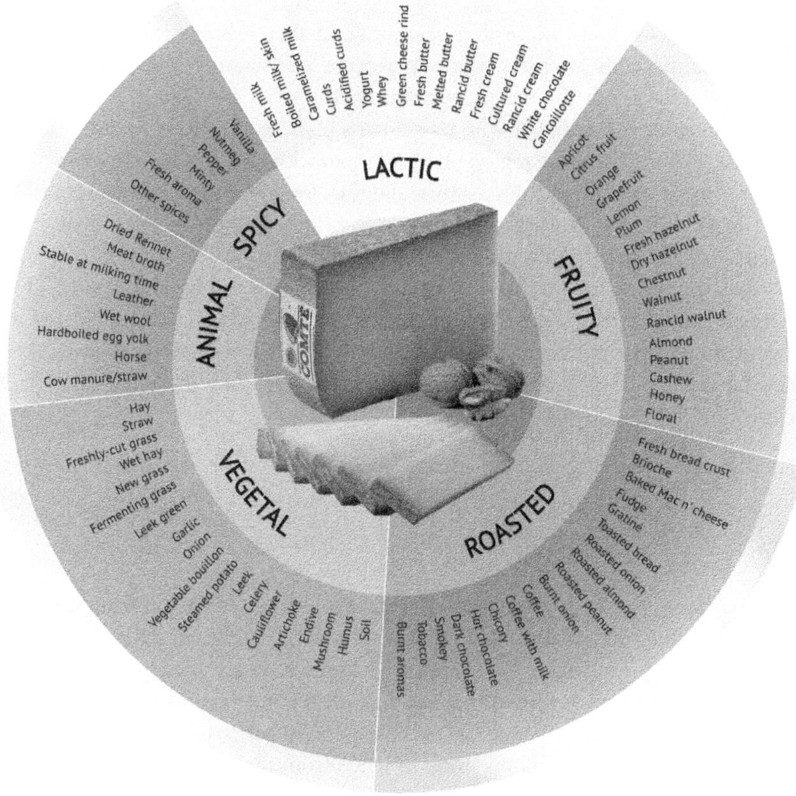

Figure 7.1 The Comté Wheel of Aromas (English translation). Copyright Comité Interprofessionnel de Gestion du Comté (CIGC).

they can also be shared. In Comté, for example, hay is ubiquitous: farmers produce their own and store it in large barns during the long winter months.

Our first point then is that while the aromatic descriptors used by the *jury terroir* function as a kind of technical language used to create the sensory characterizations that appear in scientific publications, they must also be recognized as metaphors that recall the various "domains of experience" (Fernandez 1986: 203) of the Comté chain and the wider region. For example, there are aromatic descriptors that recall production environments, such as the farm (*hay, stable at milking time*) and the *fruitière* (*curds, whey*). Other descriptors recall the natural environment: the forests (*mushrooms, soil*), the pastures (*floral, young grass*), the garden (*onion, celery*) and the orchard (*plum, chestnut*). Domestic spaces and activities are also evoked, such as the kitchen and cooking (*vegetable bouillon, hardboiled egg yolk, nutmeg*). Some aromatic descriptors recall other regional foods (*honey, cancoillote*[4]). And finally, several of the descriptors belong to more than one domain: *walnut*, for example, recalls the natural environment and other regional products (walnut is a common descriptor for the region's yellow wine).

Jury members refer to these descriptors as the "words of the *terroir*," and, as Sutton explores in his work, memory plays a key role in their recognition, circulation and emotional impact. In interviews, jury members associate descriptors with people, places and activities in the region: *hazelnut* to the fresh pressed cheese in the local *fruitière*, *horse* to a grandfather who worked as a blacksmith, and so on. Members also claim that growing up in the region provides them with a shared sensory memory that aids in learning the aromatic descriptors, and thus joining the taste community: "I was raised on a farm and have always loved to cook, so little by little the words of the *terroir* started to make sense to me" (CIGC administrator). Jury participants also recognize and enjoy the poetic nature of this language, anchored as it is both experientially and emotionally in everyday life: "Warm milk, skin of milk: it's beautiful. When you warm up your milk in the morning, and it boils over (Laughs). You know, everyone knows these terms, and they're nice. You describe a cheese, and it's beautiful" (Retired technician).

"Fresh cream, young grass, potato": An Iterative Performance of Images

While the aromatic descriptors themselves are important, so is the manner they are publically performed and shared within the frame of the *mise en commun*, which is at the heart of the proceedings. Jury members begin with a series of warm-up exercises that help them tune into one another and the task at hand. Sometimes Florence passes out a series of numbered, opaque jars and participants put words to the smells in each jar. At other times, jury members bring in items from their own sensescapes to share with the group. Following the warm-up exercises, participants turn their attention to the cheese tasting proper. This is divided into two steps, which are repeated with each cheese (and three to five cheeses are tasted per session). First, jury members taste alone, interacting with their bar of cheese and a tasting sheet. The individual portion of the tasting generally lasts around fifteen minutes and involves a series of body gestures and skills.

After the individual portion of the tasting, it is time to share and discuss. This moment is referred to as the *mise en commun*. A difficult term to translate into English, *mise en commun* means something like the sharing, pooling together or placing in a commons. Participants share their results for *odeurs* (the smells of cheese when held up to nose), the *saveurs* (or flavors in English: sweet, salty, bitter, acidic) and *arômes* (the smells of cheese when in mouth, which I refer to in this chapter as "aromas"). The *mise en commun* is highly structured and rule-governed, and consists of two phases. Consider the following example of the *mise en commun* of aromatic descriptors:

> Florence begins the *tour de table* by asking Jean,[5] a farmer and longtime jury participant sitting to her right to begin. He does so by stating: "Mushroom, curdled acidic milk, lemon, chocolate."

Next to him is a new recruit, Dominique, a farmer also. She hesitates and Florence tells her that she doesn't need to share yet if she feels uncomfortable. Dominique says it's ok and offers: "Fresh cream, young grass, potato."

And the *tour de table* continues, with no further prompting from Florence, as participants state three or four descriptors, in order, around the table. The continual repetition of aromatic descriptors assumes a certain solemn tone and rhythmic cadence:

> "Roasted hazelnut, curdled acidic milk, vegetable broth." (Luc, culinary instructor)
> "Skin of warm milk, fresh hazelnut, endive." (Philippe, technician)
> "Brioche, fresh hazelnut, citrus fruits." (Daniel, retired culinary instructor)
> "Lemon, fresh milk, milk chocolate." (Francine, retired technician)
> "Brioche, peanut, fermented grass, white chocolate." (Andre, taste educator)

And on they go around the table until everyone has spoken. Then it is time to discuss. Florence begins by saying "fresh hazelnut comes up often and there is also a fair number of brioche and warm lactics like melted butter or the skin of warm milk. But what about mentions of curdled milk? Do we take it that far?"

Discussion ensues. Some participants offer additional descriptors they wrote on their sheets but did not include in their principle aroma set:

> Victor (farmer): "I have fresh cream and brioche."
> Alain (technician): "I put *rognure* [fresh cheese that is left after pressing]. Not far from brioche, I put pasta gratin. I could have put brioche".
> Carole (taste educator): "Me too, I also put pasta gratin."
> Discussion goes on for another several minutes and green and citrus notes are also evoked. To conclude Florence offers: "if we want to summarize we could put brioche, fresh hazelnut, white chocolate, vanilla and from the fruity family we also have a bit of citrus."[6]

Novices sometimes mention that their first *mise en commun* is overwhelming, even if they, like confirmed tasters, identify it as a favorite moment. Certainly, the silence that falls upon the room and the rhythmic cadence of utterances amplifies the poetic nature of the words. The table also organizes utterances; sharing always begins at one side of the table and proceeds to the other, with little prompting from Florence. In interviews Florence stressed the importance of an open table. When she first started the jury, she used "boxes," as she was taught in school, where tasters sit in individual cubicles with blinders set up on all three sides. However, she soon moved to the "convivial table, so that participants could, in a shared space, taste in silence and then communicate: talk about their emotions, talk about what they thought about the product."

Of equal importance to the spatial organization and rhythmic cadence of utterances, is the repetition of terms: within one *mise en commun*, across the three to five *mise en commun* of one jury and then across time, each time the jury meets. Fernandez argues that rituals provide for the "performance of a sequence of images" that "revitalizes, in effect and by simple iteration, a universe of domains, an acceptable cosmology of

Figure 7.2 The *Jury Terroir*. Photo by Thierry Petit.

participation, a compelling whole" (203). Sutton applies this same idea of revitalization through repetition to the manner in which the smell of basil or the taste of feta serve to reconnect Greek expatriates to their homeland. Similarly, in this case, the solemn listing of aromatic descriptors within the *mise en commun* also acts as a "performance of images" that gathers domains together into a compelling representation of a whole, which can be understood as the *terroir*.

Fernandez also draws our attention to "the productive tension between differentiated domains, on the one hand, and their collapse into wider classification on the other" (206–207). When applied to the *mise en commun*, we see that, on the one hand, in and across these moments, the cheeses, the *fruitières* and the tasters reveal and are granted their own unique identities, while, at the same time, the repetition of named and detailed differences by way of aromatic descriptors also produces a "superordinate semantic category" (Basso 1976 cited in Fernandez 1986: 204), or a larger inclusive category that contains the others, thereby collapsing these individualistic differences into a single message, or belief. In the *jury terroir* that message is "diversity," reiterated and embodied, as the core characteristic and value of the chain and the *terroir*.

Consider, for example, the cheese. What is clearly communicated across *mise en commun*, is the notion that each wheel of Comté cheese is unique. Interestingly, during this process, the cheeses become agents in their own right. In interviews, for example, jury members refer to the cheeses as speaking subjects: "The cheeses do not always speak to us in the same way; some are more or less talkative, and, of course, our taste buds can be more or less attentive" (culinary instructor). In other words, the cheeses here each become distinctive agents that speak to the tasters through their unique aromatic profiles. At the same time, the diversity of Comté cheeses is reiterated and valued; a juror states: "the *jury terroir* is there to speak of all the richness in the taste of one Comté, in order to talk about the diversity of Comtés" (technician).

This dialectical interplay between part and whole also applies to the *fruitière*. To recognize and celebrate the uniqueness and diversity of the cheeses, is to also recognize and celebrate the uniqueness and diversity of the *fruitières*. For example, at the end of the *mise en commun* cited above, Florence announced, as she always does, the name of the *fruitière* that made the cheese we just tasted. As bits of left over cheese were collected, my table neighbors began a discussion:

> **Alain** It's the X co-op, huh? [All *fruitières* are "cooperatives" and so sometimes referred to in this way. They are also referenced by the name of the village they are located in. Here Alain names the village, though I refrain from including it in the publication.]
>
> **Sandrine** Still in the same place?
>
> **Carole** Such an amazing place, right?
>
> **Marilyne** Yea, really great.
>
> **Carole** Christy, do you know it?
>
> When I respond in the negative Carole exclaims, "You really should visit it one day (…). It is really unique, one of a kind." We turn our heads to the poster of the region's *fruitières* hanging on the wall behind us, to see if this particular co-op is pictured there. It isn't, and so they describe the road that leads to it.

Several months later, I happened to visit the small town in question. I found the road and followed it to the co-op. To some extent I already felt an attachment to this *fruitière*, and felt I had met its cheese. In other words, I had been socialized into recognizing and valuing individual *fruitières*, and, simultaneously, the system of diverse *fruitières* as a cherished part of the chain and the region. It happens still from time to time, while I am driving through the region with my family, that we chance upon a village that carries the name of a cheese I have tasted in the jury: "Oh, there is a *fruitière* in this village!," I'll exclaim, "Do you mind if we make a stop?"

However, the most surprising representation of diversity is the array of sensorial subjectivities performed and displayed by participants. In interviews, jury members underline the fact that everyone who comes to the jury is different, with their own individual tastes. As one participant (a farmer) states: "Often, during the *mise en commun* … we see that each person has a taste sensibility … and I find it really wonderful. We see the diversity of the people through their tastes." Sometimes a person's sensory subjectivity is understood to be a reflection of their profession. For example, a farmer and novice jury member explained to me that the smells from the farm were easier for him to recognize. At other times, a person's penchant for an aroma has little to do with a profession. While trying to explain this to me, for example, Bérodier notes, "Jacques, for example, was someone who found fruity notes … Daniel is a fan of vegetable … Luc finds spicy notes easily … Francine is very geared towards roasted … Almost everyone kind of has their specialty." Thus, the *mise en commun* is also a sharing of the sensorial subjectivities of chain and regional actors.[7] In this way the *jury terroir* is also a reminder that to produce a Comté, different sets of actors need to work together. A wheel of Comté is never a farmstead cheese: it is, instead, the "fruit" (hence the word *fruitière*) of a collective effort connecting, first, the farmer, the cheesemaker and the ager, and then beyond this trio, implicating other chain and regional actors as well, including *jury terroir* members.

"It's like lying in a field": On Emotion and Transcendence

The "whole," according to Fernandez, is "a state of relatedness—a kind of conviviality of experience" (191). Such feelings are commonly associated with this tasting practice; for example: "the *jury terroir* is a tasting, but also an exchange, the sharing of sensations, discussion and conviviality" (CIGC administrator). Participants also refer to a sense of unity that transcends their diversity: "We have lots of people that come from all sorts of environments, all sorts of backgrounds, but there is a kind of unanimity, a kind of symbiosis around the table that helps us come to a result, that makes things go well" (farmer). This last citation suggests that the notion of "communitas," as elaborated by Turner (1969), may also be a fitting analytical frame for the jury. Through its temporal organization in particular, the jury may very well produce a sense of a community that transcends the social hierarchies that characterize daily relations within the chain (where, for example, a technician's taste talk outweighs a farmer's). Fernandez, however, draws our attention to the manner in which rituals are in dialogue with wider contexts. In particular, he argues that the sense of wholeness experienced in such rituals must be understood as a response to the fragmentation and alienation that characterizes moments of acute change. In this way, Fernandez examines how the revitalization rituals of the Fang of Gabon respond to the alienation and fragmentation brought on by "agents of the colonial world and simply modern times" (1986: 562), while Sutton (2001) examines the manner in which foods from home assuage the estrangement and longing associated with the experience of displacement.

French farmers, too, face the alienation that can accompany processes of modernization and globalization (Rogers 2000). For example, while cooperation and solidarity among actors is central to the survival of the Comté chain, individual actors—busy with their jobs, with their families—do not always have time to reconnect with others in the chain. The jury provides a space for reconnection: "You know, I run around all day, there is so much to do on a farm, so much to do, in fact, that I rarely do anything else, but when I come to the jury, and I always have to hurry to be on time, it's a moment to be with others, to get out of my little bubble." Getting out of "one's bubble" provides for an emotional experience of reconnection, but also plays an important political role within the Comté cooperative system, where actors—within the *fruitière* cooperatives, or within the electoral colleges of the CIGC—must continually work and organize collectively, taking into account and valuing the work of each link in the chain. The jury plays such a role, as do many other activities, events and publications. The "sirens of modernity," one of the jury members likes to tell my students when we visit his farm and *fruitière* in the early months of spring, continually sing the promises of *individualisme*, but members of the chain must remain collectively organized and *solidaire*.[8]

Finally, more than a sense of connection with others in the chain, tasting in this manner also provides for a transcendental connection to nature itself. I asked a soon-to-be retired farmer, who had only just begun tasting with the *jury terroir*, why he chose to take up this particular activity. He responded in the following way:

> To look up close! Like when you lie down in a field of grass and look at the earth. It's amazing what you see! You see things that you never noticed before ... *Mais oui*! Some people don't know that a field is teeming with life. They just don't pay any attention. And yet, that is part of the existential chain of life ... And, you know, we are not even really at the top of the chain, because we are fragile, somewhere in the middle [Pause]. So, it's about returning to these things. You know, there are generations of farmers who never bent over in their fields to wonder at all the life that abounds there.

This was one of the last interviews I conducted and I find it rather fitting that my quest to understand how this tasting activity acts to create and sustain a belief in *terroir*, comes full circle, to end with a metaphor that returns us to the earth (*la terre*, which in French refers to both "earth" and "soil"). Tasting in the jury, like lying down in a field and closely observing the life teeming there, is a sensorial experience focused on the minutiae of everyday life that gives way to a higher order reflection and sense of connection. This suggests that the *jury terroir*, and perhaps the *mise en commun* of aromatic descriptors in particular, is cosmological in another sense, in that it invites us to wonder at the wider universe, and the natural world in particular, as well as our place within it, if not our dependence upon it. We also note that the farmer here refers to a "return" to a relation with nature that is not readily available in the world today, even for farmers. Fernandez identifies this as the "'shock of recognition' of a wider integrity of things" (205). Shock here implies a sharp and sudden feeling (an epiphany perhaps?), provoked by "the collapse of separation into relatedness ... the recognition of a greater whole" (205). In this way, through the "figurative language and the argument of images" (206) involved in the exchange of aromatic descriptors, the jury members achieve "a wider and more transcendent view of things" (206).

"We are actors!": Concluding Remarks

Let's return to taste. The concrete elaboration of a shared belief in *terroir*, and thus of a strong collective identity, happens through the mobilization of one of the most subjective aspects of our relationship to cheese: taste. The *jury terroir* is a space for the articulation of an individual and collective taste, or, more precisely, for the articulation of individual subjectivity with(in) a larger intersubjective whole.

In this way we suggest that tasting be viewed here as an important form of agency. Taste is often thought of as the language and practice of the connoisseur or the affluent consumer, and thus as a practice of distinction or domination. As many scholars have shown us, there is power in this language, the practices attached to it, and the people who possess it. However, when it is mastered and led by the actors of a supply chain and a region, indeed developed by and embedded in their worldview, there is a bit of power to be had there too. And empowerment, especially of farmers, has always been an important goal of the *jury terroir*. "Farmers live in enormously rich sensory worlds and have amazing sensory memories," Florence told me, "they just hadn't ever put that into words."

Of course, this does not mean that the notion of taste as elitist has disappeared, nor that all Comté farmers choose to practice taste in this way. Indeed, the CIGC's early dream of generalizing this tasting practice throughout the chain did not come to pass. As Jean-Jacques Bret, director of the CIGC from 1983 to 2013, states:

> If we didn't generalize it more, it is because it is too sophisticated, in my opinion … They have an empirical knowledge of their product … they do not necessarily feel the need to define it with these exact terms … When you eat a piece of Comté and, after a certain number of experiences at the *jury terroir*, you are able to say that it smells like dark chocolate or brioche (pause) that irritates people … You know, when I drink wine, I don't bother with knowing if it smells like banana … I think: 'Shit, this is good!' *Et voilà*! I am like any average French person. In the act of daily consumption, we don't bother.

So interacting with the cheese in this way is not meaningful to everyone, which raises important questions about tasting and the taste amateur that will need to be addressed in the larger research project. However, even if not everyone in the chain interacts with the cheese in this way, the *jury terroir* remains an essential collective activity for two reasons. First, because it provides a language and practice that has convinced others within the chain, and without, that diversity is a defining feature of the chain's identity and highly valuable. As Jean-Jacques Bret says: "The work of the *jury terroir* allowed the chain to lay claim to its identity, to better appropriate its own values. We are an artisanal cheese and so the tastes are different, and we are capable of being proud of this because we are capable of describing them." Relatedly, and this is the second reason: the *jury terroir*'s tasting practice incarnates and reiterates the centrality of both individual and collective agency within the chain. Being an actor (*acteur*), or an agent in their own destiny, Bret told me, is what separates Comté farmers from others:

> When there was the mad cow crisis, it was surprising, we saw farmers interviewed on TV that were crushed by the system: they didn't eat their own animals because they didn't know what was in the feed. They were crushed by the banks. When you interview a guy here, the difference is impressive. These people are actors. They yell, they are not happy because we took this step or that, but that proves that they are actors. They are actors because they define what is in the production rules, they are actors because they run their *fruitières*, they manage their environment, they are involved in their communities … There is a sort of global logic, a global approach to their environment that makes these people actors. They are actors in the organization of their cheese cooperatives, they are actors in the identity they give to their product.

Here, Bret reminds us that the interactional practices at the heart of the *jury terroir* echo the interactional practices at the heart of the larger cooperative structure of the Comté chain where, for example, farmers own and run the cheese-making facilities and benefit from a strict profit-sharing system. In this way, the *mise en commun* is a ritual that symbolically incarnates and supports, indeed even brings jury members to embody, the larger political-economic model of the chain, the two recalling and reinforcing one another. Inversely, we also understand through this example that "embeddedness" is not only a question of institutions, regulations and shared beliefs, but also practices, mundane and ritualistic, that provide individuals with voice and reconnection in a manner that works against the alienating and isolating forces of modernity.

Notes

1 The *Comité Interprofessionnel de Gestion du Comté* regulates the Comté label and codifies the rules of production.
2 See Sutton 2010.
3 The jury's current list of descriptors includes some new terms that are not on the wheel.
4 *Cancoillote* is a local cheese produced in the Haute-Saône department.
5 I have changed participants' names for publication.
6 The final publication of the *fruitière*'s aromatic profile is constructed by Florence using statistical analysis drawn from seven to ten separate tastings. Novices' results are not included in the tally of final results.
7 An equally important part of the work is then moving towards one another during the discussion; an activity framed by a set of participant rules that we do not discuss here.
8 See, for example "Scénario 'individualisme': Quand chacun tire la couverture à soi, tout le monde a froid" (The Individualism Scenario: When everyone is trying to hog the blanket, we all end up being cold). Les Nouvelles du Comté, No. 96, Hiver 2017.

References

Barham, E. (2003). "Translating Terroir: The Global Challenge of French AOC Labeling." *Journal of Rural Studies*, 19: 127–138.
K. Basso (1976). "Wise Words of the Western Apache," in K. Basso and H. Selby, eds., *Meaning in Anthropology*, 93–121. Albuquerque: University of New Mexico Press.
Bérodier, F., C. Stevenot, P. Schlich (1997). "Description of the Flavor of Comté Cheese." *Food Science and Technology—Lebensmittel-Wissenschaft & Technologie*, 30: 298–304.
Bowen, S. (2011). "The Importance of Place: Re-territorialising Embeddedness." *Sociologia Ruralis*, 51(4): 325–348.
Feld, S. and Basso, K., eds. (1996). *Senses of Place*. School of American Research Press.
Fernandez, J. (1986). *Persuasions and Performances: The Play of Tropes in Culture*. Bloomington, IN: Indiana University Press.
Hennion, A. and C. Teil (2004). "Discovering Quality or Performing Taste? A Sociology of the Amateur," in M. Harvey, A, McMeekin, A. Warde, eds., *Qualities of Food*, 19–37. Manchester: Manchester University Press.

Leynse, W. (2006). "Journeys Through 'Ingestible Topography': Socializing the 'Situated Eater' in France," in T. Wilson, ed., *European Studies*, 22: 129–158.

Monnet J.C., F. Bérodier, P.M. Badot (2000). "Characterization and Localisation of a Cheese Georegion Using Edaphic Criteria (Jura Mountains, France)." *Journal of Dairy Science*, 83(8): 1692–1704.

Paxson, H. (2012). *The Life of Cheese: Crafting Food and Value in America*. Berkeley: University of California Press.

Rogers, S.C. (2000). "Farming Visions: Agriculture in French Culture." *French Politics, Culture and Society*, 18(1): 50–70.

Rosaldo, R. (1993). "Grief and a Headhunter's Rage," in *Culture and Truth: The Remaking of Social Analysis*. Boston: Beacon Press.

Shields-Argelès, C. (2016). "The Comté Aroma Wheel: History of an Invention, Ethnography of a Practice. A Look at the Early Years," in. M. McWilliams, ed., *Food and Communication*. London: Prospect Books.

Sutton, D. (2001). *Remembrance of Repasts: An Anthropology of Food and Memory*. New York: Berg.

Sutton, D. (2010). "Food and the Senses." *Annual Review of Anthropology* 39: 209–223.

Trubek, A. (2008). *The Taste of Place: A Cultural Journey into Terroir*. Berkeley: University of California Press.

Turner, V. (1969). *The Ritual Process: Structure and Anti-Structure*. Ithaca, NY: Cornell University Press.

Part Three

Taste Education and Sharing: Identity and Community

8

"Listen! We Made These Potatoes Crispy!"

Danish Adolescents Sharing Taste in a School Class

Susanne Højlund[1]

Introduction

In 2015 I was visiting a school class in a suburb of one of the larger cities in Denmark, interviewing the children in fifth grade about their lunch preferences. Three girls were sitting in a group and eating pasta from the local school cafeteria. On the table six pieces of bacon were lined up waiting to be eaten. The girls explained to me: "We are the bacon class. We simply love bacon, and often order this pasta salad because there is bacon on top." "But why don't you eat it?" I asked them, referring to the nicely ordered bacon slices in front of me. One of the girls answered: "Because we like to wait until we are all ready, and then eat them together, like a kind of dessert."

The girls' alliance on bacon made me curious. As did another example from my fieldwork at a food festival, where we asked children to find things they liked and disliked and report to us what they were (Højlund 2015). All of the thirty-four participating children wanted to solve the task with a friend, often a classmate, and they returned to us with answers such as: "We liked the cheese in the tent over there", or: "We liked the fish balls, but we didn't like the ice cream with insects . . .". Not one of the children reported an individual taste preference to us, but all told us in one sentence what they both liked. As one nine-year-old girl expressed it: "We couldn't find something we disagreed on because we have nearly similar taste buds."[2] (Markvorsen 2015: 24).

As my fieldwork at different Danish schools during 2015 and 2016 developed I began to reflect on the special "we" that I often met among the children when talking with them about food and taste preferences. The general idea in Denmark is that children—except for a stated craving for sugar—have their individual perceptions of taste, and are often picky eaters giving their parents trouble around the table exactly because of this individual approach to taste (DeCosta et al. 2017). During my fieldwork, another hypothesis was taking shape as I talked to more and more children about their experiences as tasters. I became aware that children's likes and dislikes are neither stable conditions, nor individual—they are neither picky, nor not-picky. A boy for example told me, that he liked red pepper at his grandmothers' house, but not at home.

Another reported that he never ate white bread at school, but loved it in the after-school-institution. The social context seemingly plays an intriguing role in children's definitions and experiences of taste (see also Torralba and Guidalli 2015).

So, what is taste then? Certainly, more than an individual preference; not simply a cultural determination either. My idea is that there are lessons to learn from exploring how children experience and share taste in their everyday institutional lives. This approach makes it possible to understand better how taste can be seen as a social phenomenon, how taste can be contagious and spread from the "I" to the "We." This chapter builds on fieldwork done in 2015–2016 when I followed two cooking classes at a rural school in Denmark (fifth and sixth grade home economics classes), and in addition made shorter visits to four other urban schools' home economics classes. I also refer to children participating in a local food festival, where I set up anthropological experiments every summer from 2012 to 2016 (Højlund 2015), and the experimental project Cook & Cut.[3] I visited the classes once or twice every week, cooked with the children, interviewed them about their interpretations of taste, ate lunch with them and interviewed their teachers. The children also made films about their cooking and drawings about their food habits, and I recorded sound and took a lot of photos of them working with food. My project to explore and communicate how children create knowledge on taste is part of a national research and communication center named "Taste for Life" (2014–2018).[4]

In this chapter I analyze how young pupils' (eleven to fourteen years old) taste experiences and sociality are intertwined. With examples from their interactions during home economics lessons I show how they use taste strategically to create and re-create certain social dynamics. I show how taste can be used to include and exclude others, to create collective work, to perform identity, and to generate an "us" as well as a "them." The chapter is thus an attempt to further develop my argument that taste is a social activity (Højlund 2016). How taste is learned, shared, adapted, and communicated in everyday interactions is still rather unexplored, and the examples here just begin to shed light on what Chau has called "the sensorial production of the social" (2008).

Danish Children and the Politics of Taste

Danish children attend public school from six years old; they normally continue at the same school until they are sixteen years old, and they often even stay in the same class and have the same teachers. Different schools plan their curriculum differently, but in general pupils are offered lessons in home economics in fifth and sixth grade with two to three lessons per week.

The concept of taste has recently become part of the national school curriculum, obliging teachers to put taste on the agenda in the classroom; but also in Danish society in general children's taste knowledge has generated increasing interest. Famous chefs are organizing events with the purpose of "nudging" children to taste new food, to gain what they name "taste courage."[5] TV programs for children and children's cookbooks celebrate children's experimental cooking and the child chef (Leer 2015).

A lot of Danish applied research projects focus on how to learn to taste. An example is the Danish "mass experiment" where around 20,000 children were given a "taste kit" and a questionnaire about their taste preferences. The results and the interpretations hit national news programs and produced many articles not only in scientific journals, but also in popular media. The headline was that "children can learn to eat everything." No excuse for not serving broccoli anymore, it's all about recognizing the taste. Children get their taste palate from local food traditions and family food culture (Petersen 2015). The common idea among parents—that children just have to taste food they don't like fifteen to twenty times and then they will like it—was confirmed through this mass experiment. The preoccupation with children's neophobia has a long tradition in nutritional research (see e.g. Birch 1999, Lynch 2010). Exposure frequency is the magic code (Birch 1999). Nudging is another (DeCosta et al. 2017: 341).[6]

From this perspective children's taste is already part of the public discourse. Experts tell us that newborns have a natural, biological preference for sweetness, that children do not reflect on what is healthy but only act on their spontaneous taste attractions and longing for pleasure, and that they are often afraid of new foods, and therefore often are picky eaters (Science Daily 2013).[7] These arguments are repeated in books and magazines for parents problematizing children's tasting practices, and representing them as passive recipients waiting to become reasonable and civilized adults being able to control their desire, to reflect on their choices and to expand their taste preferences.

The main focus of child-and-taste initiatives in Denmark is treating taste as an imperative, and as a pedagogical tool. Learn to taste! Challenge your taste buds! Teach your children to taste! Educate the pupils in taste! In particular, the taste of sweetness has been an item on political agendas. The argument that children have a special sweet tooth, combined with arguments from nutritional research, food ideologies and health campaigns—has created a strong "discourse cocktail." This has e.g. paved the way for the special Danish "sugar policy" in children's institutions, where there is a recommendation not to serve sweet beverages.[8] Nearly all schools and kindergartens in Denmark have formulated limits on how many cakes and sweets the children can eat in the institution (Nielsen 2012). Such limits are represented as good for their health and also a means of civilizing their sense of taste, communicating the idea that sweetness can be dangerous.

In the specific Danish context children's taste buds are thus an issue of political interest as never before. Regardless of whether you sympathize with this agenda it does not give much voice to the children themselves. The research that informs the public discourse on "the child eater" mostly works with nutritional, psychological and consumer studies approaches with no scientific tradition of considering children's knowledge as valid. Actually, many arguments about taste preferences are built on experiments with rats (Birch 1999). Within these research paradigms children are seen as vulnerable (easy to spoil—by media, market, peers and bad parenting), at risk (of obesity), as problematic eaters (suffering from pickiness and neophobia) and as more or less passive respondents (to taste stimuli).

In spite of the last twenty years' development of a sociological/anthropological research paradigm of "the competent child" (for an overview see: James 2007) and the derived empowerment of the "child cook" (Leer 2015), there is still a general absence of children's voices in research on children's relations to food (Johansson et al. 2009). The dominating political concept of taste is about the developing child, being the object of a moral (adult) agenda. Even though children are represented in TV-shows, experiments and interviews as never before, and there is a huge amount of research on children's nutrition and their relation to healthy and unhealthy food, we still do not have a solid, qualitative research platform which addresses questions of how children interpret taste, experience their tasting, and negotiate and share their preferences (Leer and Wistoft 2015).

The Social Activity of Tasting

In order to create knowledge about how children generate and share interpretations of taste it is necessary to reflect on the notion of "taste" as an analytical concept. The common-sense understanding of taste as an individual perception, happening in the brain of the eater, or as something inherent in food, needs rethinking. This approach mostly conceptualizes taste as an individual response or assessment (Teil and Hennion 2004). Taste is essentialized and seldom investigated as a cultural construction shaped within our social relations. As early as 1989, Paul Stoller invited anthropologists to do "tasteful ethnography" (Stoller 1989: 27) focusing on people's strategic use of their senses. However, there seem to be few anthropological attempts to theorize the senses in relation to "the social," although David Howes (2005, 2014) has explored how our senses are culturally shaped. But how these "ways of sensing" (ibid.) are constructed in relation to everyday interactions and practices is seldom the object of fieldwork interests. As Adam Chau states:

> Yet something seems to be amiss in these often elegantly constructed ethnographies of the senses: the role of the social, and not only how the senses receive and perceive the social but more importantly how social actors actively construct their social worlds in sensorially rich manners, and how moments of sensorialized sociality become institutionalized.
>
> Chau 2008: 487–488

An exception is Stoller's analysis "The Etiology of bad Sauce" (Stoller 1989: 19). By paying attention to how one of his informants in certain social contexts cooked a very bad-tasting sauce while in others was able to produce a very good-tasting one, Stoller got access to valuable knowledge about this informant's social position in her local community and conceptualized taste as an activity. This is in line with Chau (2008) who advocates for a focus on production instead of interpretation:

> Going beyond these sensory interpretations, however, a sensory-production approach would instead focus on *the production side* of the participating agents.
>
> Chau 2008: 490

Chau is pointing to the "doing" as a co-production of a sensorial rich atmosphere, what he names a *social sensorium* (ibid: 489).

The lessons in home economics represent such a sensuous context. At the beginning of the lessons the school kitchen was a prototype of a school classroom. Pupils were sitting at a table, the teacher was standing and giving instructions, and the children were asking questions with hands up. Half an hour later the scene had changed. The children were now standing, working in four different groups, each group around a stove. Some of them were washing and peeling potatoes, others cutting onions, some roasting meat, others searching their iPad to find a recipe. Most of the children were talking, standing in groups, looking at the food and commenting on their own work and that of others. The teacher was walking around in the background, cleaning shelves, giving advice here and there, but not interfering much. As time passed, the sensory intensity increased in the kitchen. The scent of food filled the room, more and more sounds were produced: of laughter and voices, of water running, of kitchen equipment being moved around. Menus were produced and watched, and serving arranged with artistic engagement. The kitchen became like a beehive, buzzing with activity.

The kitchen atmosphere is generated by the sensuous work of people in the room. It does not come from the outside. The children are co-producers of a kind of "red-hot-sociality" (Chau: 2008) where all senses are in play and create an atmosphere that all in the room are part—and co-producers—of. With this approach taste becomes a sensuous activity. Taste is not a fact, but a process of *tasting activities* generated by a relation between food and subject (Teil and Hennion 2004). Teil and Hennion stress that this happens in "taste collectives" (ibid: 24). I add to this point, that not only is collectivity creating taste, the activity of tasting itself creates collectivity.

Bringing Taste to the Public Sphere of the Classroom

The following examples from my fieldwork[9] are chosen to illustrate a variety of tasting activities—or ways of tasting—and how these provide different social dynamics in the class. During my analyses of these examples I have not looked for the children's specific definitions of taste. As is the case in society in general, taste had many meanings for the pupils. A quality of a food product, a multi- or a mono-sensory experience, a tongue or a brain response to a stimuli, a lifestyle etc. The children picked up these different meanings of taste and put them in play in their social relations. With Tuomela's distinction between "I-mode" and "We-mode" (Tuomela 2007), I have looked for examples where the children enact tasting and thereby create a group ethos, in an interplay between the private "I" and the public "We."

Taste my Tea! Sensory Invitations

One early morning we were silently waiting outside the school kitchen for the teacher to arrive and open the locked door for the next two-hour class. The children arrived in ones and twos. Sarah joined the waiting group carrying a cup which she held with both hands close to her mouth. She sipped the steaming fluid and started talking about her new tea that she brought from home. She invited the two girls around her to taste it, while she commented on the blueberry taste: "We have got a new tea at home—do you want to try it? It tastes of berries. I really like it." Two girls standing around her tasted the hot tea and answered with a "mmm." Another girl, seemingly inspired by the conversation about taste, said that she had bought a packet of banana tasting chewing gum which she really didn't like, and she asked if anyone wanted to have the rest of the packet. Before anyone responded the teacher arrived and the children—now very lively and talking—picked up chairs and brought them to the room, where they sat down around four tables.

This example represents a basic strategy for creating a collective awareness, or "we-ness" around taste. Sarah invites some of her classmates into her private sphere of taste and thereby brings the "I" to the "we." By accepting her invitation these girls suddenly have a small taste community, a common experience, that generates a broader debate on the taste of chewing gum.

I call this a "sensory invitation" and not a "taste invitation" as the sharing is multisensory: the girls look at the tea, they smell it and touch the warm cup. I see it as a basic strategy for sociability—inviting others into one's sensory experiences is a very common action. Human beings are used to sharing sensory impressions by inviting others not only to taste, but also to see, touch and smell (Pink 2007). Taste has this ability to create a "common third" (Husen 1996). This is a topic of common interest and engagement, which the following examples will also show.

Listen, we Made these Potatoes Crispy! Proclaiming Sensory Skills

The class loved the days when the teacher brought different foodstuffs to the table and asked the children to invent menus from the materials. One of these days Jens and two other boys had chosen to make potato crisps. They cut the potatoes—many of them—in very thin slices and fried them in a hot pan. They were all very engaged in the process of cutting, roasting, and drying the potatoes on paper, then adding salt. The boys seemingly had high expectations of the result, and received a lot of attention from their classmates. When the moment came to taste the result, all three boys stopped working while Jens slowly took one of the crisps up to his mouth, and crunched it between his teeth. He then proclaimed loudly and excitedly: "Listen! We made these potatoes crispy!"

There is a lot at stake in this statement. First, Jens was proud of the result, and wanted to tell people about it—his tasting was made public. Crispiness is an ideal for crisps (Spence 2015), and shows that the boys were working towards an ideal taste experience: the crispy-ness. Secondly, Jens demonstrated in his statement, that he interpreted taste as a multisensory experience. It is the sound, he was referring to, as an important part of the taste. But he was not only referring to himself, he was referring to a "we." The taste result was a product of the group.

Embodied knowledge about taste is at play, about kitchen skills, or what Ulloa names *sensory skills* (Ulloa 2017). In her study of chefs working together in a Spanish restaurant she stresses the importance of the chefs working together. Their knowledge of taste can only be validated and developed in a social context and this sensibility of taste only develops when it is put into use together with others. Jens' statement was a celebration of the work of the group. This communality of taste, developed during the children's common taste production, was constantly in play in the classroom. But not always as loudly proclaimed as in this example.

The Work has to be Done! Tasting with your Fingers

The hands and the sensing of the fingers play an often overlooked role in the process of tasting (Mann et al. 2011). When reflecting on my fieldwork notes and photographs I became aware of the many images I have of children's hands working with food: peeling, stirring, washing, cutting, kneading, handling different utensils, decorating cakes, and putting things in order on a pan. It is interesting, though, that all these images of working hands are pictures of many hands, not just one pair of hands. It shows children working together on a process or a product. Often two or three children were standing around a pot cleaning and cutting vegetables talking about things other than the food, e.g. the forthcoming Christmas or an event at the school. Their hands were smoothly coordinated giving space to each other. These "silent producers" were not making taste public through language or performance, they just did what needed to be done. But the togetherness through routine work and craftsmanship (Sennet 2012) made them share taste at another level. The visuality, embodiment, and communication through gesture while handling food production (ibid: 207) produced a pre-taste (Mennell 1996) and a possibility to learn tacitly from each other.

We Need More Red on this Cake! Collective Creativity

The hands were also in the centre in this example, but here not occupied with routine work. The children were given a task and some ingredients and we then followed how they solved our challenge, e.g. to make a dessert from eggs, sugar, fruit, cream and sweet bread. All the groups were very engaged with the production of cakes and desserts, and in particular the decorations were hotly discussed and handled with accuracy and creativity. Often many children were involved at the beginning of the work process, exchanging ideas as to how the perfect cake should look and taste, but as time passed only the very engaged were left, perhaps due to too many discussions of which way to go, and whose idea to follow. This last process took time, as when Leo and Dina were standing around a green and red cake for half an hour discussing coloring, and decorating with sugar and berries. "We need more red on this cake!" Dina stated, and Leo followed her idea and spread more red sugar with small movements and a precise eye. They had developed an idea of decorating with the name of the class on top, and distributing red and green sugar on top making it form "6 D" in a sugar pattern. The example shows how joy and engagement but also conflicts are part of the creative process: from indeterminacy to a common product. In this case taste generates a common occupation (Lawler 2003) and expectations of the forthcoming tasting.

Figure 8.1 Sharing taste through hands working together. Photo by Susanne Højlund.

Figure 8.2 Creating taste through collaborative creativity. Photo by Susanne Højlund.

I Have to Taste what we Made! Synchronizing Taste Buds

One day the class was asked to make meatballs. I had chosen that day to ask the children for an individual interview (in another room), where I asked them about their personal experiences with taste. One of the boys explained to me that he hated minced meat so much that he wouldn't put it into his mouth at all. I replied: "Then you are not going to taste the meatballs today?" He looked at me in surprise: "Yes, of course! I was part of making it, so I have to know how it ended up tasting."

The boy's reflection exemplifies how he is able to put aside his own feeling of disgust in order to maintain his attachment to the group. Taste is the final product of their work and endeavors, it is what finishes the process. Without participating the whole way through a work process you cannot maintain your connection to the group (Husen 1996). The boy thus had to synchronize his taste buds (Brembeck 2009) with the others in the group in order to stay connected.

Do you Dare Taste That? Performing Transgression

During the home economics lessons many secondary activities took place. Due to the characteristic organization of the classroom activities, pupils could easily walk around doing other things than just being focused on food making. One example of a "side activity" also involved the act of tasting. Two boys had found a pepper container and were giving each other the challenge of tasting as much as possible. This activity was perhaps stimulated by an ongoing Danish TV-series where chili tasting is the central activity. Different famous people are challenged to taste very strong chili and the camera simply shows their reactions. Disgust and courage are at play here, and individual performance based on tasting is the central plot. The "we-ness" of this tasting activity is created through the entertainment effect that generates an audience of other children watching and laughing. To taste is also to perform, and performance demands an audience (Teil and Hennion 2004: 32). The tasting "I" has the central role, communicating to the "we" how transgression is experienced.

Only we Have Access to this Taste! Distributing Access

Penny van Esterik has developed the concept of *commensal circles* in order to shed light on the many different ways we socialize around food (van Esterik 2017). The school children are expected to know these rules for eating together around a table, and sharing the food. Every home economics lesson ended with the children eating in groups around four tables. The children were obliged to set the table themselves and were meant to share the food they had produced in the group. Often the different groups showed a lot of interest in the others' products, especially if they were popular dishes. During a lesson the children were asked to make food they remembered from a holiday with their families. One group chose to make cheese-nuggets and were very engaged in counting the amount they produced. Again and again a girl repeated how many there were now for each of them in the group. The teacher asked laughingly: "What about me then? Am I not supposed to taste?" She was then allowed to taste one, while the four children in the group shared the rest, with the class observing them with envy.

As resources were limited in the school kitchen, it was not possible to eat around a big table and let everybody taste. The children had to develop skills as to how to distribute access to the desired tasting, and the logic was often that the right was given to those who had been part of the taste production. The position as distributer, defending the group borders, was tied to the social hierarchy of the group. Often the rules were accepted by all, but examples of protest and quarelling about the justice of how to share were also common.

The Wider Context of Tasting Practices

Taste—in its many forms—makes it possible for the children to create varying social platforms for activities of tasting. Many of these platforms are not the children's own work—the school structure, the class, the time schedule, the economy, the curriculum etc. are structural factors in children's lives. The children act within this structural context, as e.g. in the following example, where time decides how the tasting of freshly baked buns was organized.

"10–9–8–7–6–5–4–3–2–1–0. Time is up and your activities must stop now." The teacher shouted to everybody, with a glimmer in her eye, as it was obvious that she referred to a popular TV-program. Here participants were competing in baking skills and the program presenter uses the same counting down as a signal for everybody to keep the deadline and stop at the same time. With this very explicit announcement the teacher underscored what is always at stake in the classroom, namely the fact that the home economics lessons are part of a time-structured school day. This premise had consequences for the sociality of tasting too, e.g. two girls had to stop baking buns even though the buns were not ready. This prompted a lot of comments from classmates while tasting the nearly raw dough, and everybody wanted to try it.

The children's engagement in tastings also related to their family life and lifestyle, as illustrated in the example of Karen who was preparing dough for a cake and went to the table where the teacher had placed different ingredients. She took the eggs and observed the packaging. Then she declared indignantly to everyone near her: "Our teacher has bought eggs that are produced from chickens in a cage." She was seemingly indignant, and shared her opinion in public. The teacher had to defend herself with an argument about the limited budget available to her. Taste as lifestyle is a matter of habitus and capital, of: "the propensity and capacity to appropriate (materially or symbolically) a given class of classified, classifying objects or practices ..." (Bourdieu 1984:173). With this announcement Karen demonstrated her capacity (or cultural capital) not only in relation to eggs, but also to the current debate in Danish society on industrialized food, ecological food products, and animal welfare.

The example reminds us that children in the classroom are not similar or equal. They each come with different life stories and lifestyles. Through my participant observation I learned that the class was made up of pupils from farmers' families in the countryside, and from families with parents driving to town every day to work; from families with many siblings, and from families with a high culinary capital (Naccarato and Lebesco 2012) and families that had no special interest in food. There were children with a lot of travel memories and experiences as food tourists, and children who had not travelled outside Denmark. This fieldwork did not allow for an extended study of the different backgrounds the children brought with them, but these probably contribute to different practices of tasting. In future research, adding this dimension will help further explain their strategies.

Conclusion: Tasting between "I" and "We."

As we have seen, practices of sensuous sharing and distribution of aesthetic knowledge about taste can be both a social glue and a disrupter. The strategies mentioned here: sensory invitations, proclaiming sensory skills, handling taste together, collective creativity, synchronizing taste buds, performing transgression, distributing access, are all different combinations of the "I" and the "We." *Inviting someone to taste* your home tea, is a way of sharing your personal preferences, as when children share their lunch packets with their peers (Andersen 2015), the "I" transforms to "we" for a short moment. The practice of *synchronizing taste buds* turns the other way round: the individual seeks to fit into an imagined "group taste" in order to be included into the "we." There is no direct invitation here, but a non-verbal expectation. When practicing *collective creativity* the common product and work process is in the center. Different "I" positions struggle to create a result that they can agree upon—they are working to create a "we"-taste. There is no pre-defined version to mirror; the final product depends on the group's ability to generate ideas together. The example of proclaiming your *sensory skills* in public represents another type of sociality. Here the group does not strive to create something "new," but to get as close to the culturally accepted criterion of crispyness as possible. There is a standard here, the group knows that, and all work for the same goal. The proclamation of success seemed to be a spontaneous action, not expressed to defend the group against other groups or individuals—just to share the excitement with the general "class-we." In opposition to this, the strategy of *distributing access* is more conflictual, defending the group's right to taste by excluding others, and thereby strengthening the borders around the group, to keep others away from tasting. In the example, *performing transgression* tasting becomes an entertaining event, inviting others to be an audience, whereas *handling taste together* is an implicit tasting activity of the silent producers.

These examples of how children bring taste and tasting to "the common third" (Husen 1996) through their actions make it possible for them to engage socially. It involves group harmony, competition and conflict. The social activities of tasting create a "Them" and an "Us" based on different combinations of "I" and "We." Although the examples are few, a single class in Denmark, and the analyses micro oriented, investigating the sociality of tasting and the role of taste for sociality can perhaps generate new knowledge—not just about children. How we as human beings create, maintain and disrupt collectivity through our everyday activities of tasting is a general question that can perhaps shed new light on our relation to food and eating.

Notes

1 I would like to thank the children and their teachers for sharing their knowledge with me. Also a warm thanks to my constructive readers Karen Klitgaard Povlsen, Jonatan Leer and Stinne Gunder Krogager. A special thanks to Carole for the many inspiring dialogues around this project.
2 Nine-year-old girl interviewed at the food festival experiment.
3 Three home economics classes were asked to invent their own menus, and at the same time make films about the process. For a description, see Gunder et al. (forthcoming 2018).
4 Taste for Life, a communication Centre funded by Nordeafonden, see www.taste-for-life.org.
5 See www.meltingpotfoundation.dk/aktiviteter/madmodsloebet (accessed October 26, 2017).
6 For a review of existing research on children's eating behavior, see DeCosta et al. 2017.
7 See www.sciencedaily.com/releases/2013/07/130716115718.htm (accessed October 26, 2017).
8 See https://ec.europa.eu/jrc/sites/jrcsh/files/jrc-school-food-policy-factsheet-denmark_en.pdf (accessed October 26, 2017).
9 I have kept the focus on what happens between the children; what Suremain (2006, cited in Torralba and Giudalli 2015) names a *horizontal mode,* contrasting a *vertical mode* where the focus is on adult-children relations.

References

Andersen, S., L. Holm, and C. Baart. (2015). "School Meal Sociality or Lunch Pack Individualism? Using an Intervention Study to Compare the Social Impacts of School Meals and Packed Lunches from Home." *Social Science Information,* 54(3): 394–416.

Birch, L.L. (1999). "Development of Food Preferences." *Annual Review of Nutrition,* 19: 41–62.

Bourdieu, P. (1984). *Distinction A Social Critique of the Judgement of Taste.* Cambridge, Massachusetts: Harvard University Press.

Brembeck, H. (2009). "Children's Becoming in Frontiering Foodscapes." in A. James, A.T. Kjørholt and V. Tingstad, eds., *Children, Food and Identity in Everyday Life,* New York: Palgrave Macmillan.

Chau, A. (2008). "The Sensorial Production of the Social." *Ethnos,* 73: (4) 485–504.

DeCosta, P. et al. (2017). "Changing Children's Eating Behavior—A Review of Experimental Research." *Appetite* 113: 327–357.

Gunder, S., K. Klitgaard Povlsen, S. Højlund and J. Leer. (2018). In, J.S. Borgen and E. Eriksen Ødegaard, *New Nordic Childhoods: Innovative Research Designs for New Knowledge and Understandings.* Cappelen Damm Nordic: Oslo.

Højlund, S., ed. (2015). *Smagsmysteriet. Smagen under lup.* Smag for Livet: Smag #03.

Højlund, S. (2016). "Taste as a Social Sense. Rethinking Taste as a Cultural Activity." *Flavour,* 4 (6).

Howes, D., ed., (2005). *Empire of the Senses. The Sensual Culture Reader.* Oxford, New York: Berg.

Howes, D. and C. Classen. (2014). *Ways of Sensing: Understanding the Senses in Society.* New York: Routledge.

Husen, M. (1996). "Det fælles tredje—om fællesskab og værdier i det pædagogiske arbejde." in B. Pécseli, ed., *Kultur & pædagogik*. København: Hans Reitzels Forlag.
James, A. (2007). "Giving Voice to Children's Voices: Practices and Problems, Pitfalls and Potentials." *American Anthropologist*, 109(2): 261–272.
Johansson, B. et al. (2009). "Nordic Children's Foodscapes." *Food, Culture & Society*, 12(1): 25–51.
Lawler, M.C. (2003). "The Significance of Being Occupied: The Social Construction of Childhood Occupations." *American Journal of Occupational Therapy Volume* 57(4): 424–434.
Leer, J. (forthcoming). "What's Cooking Boys? The Cooking Boy in Danish Cookbooks for Children since 1975." Unpublished conference paper, 2015. Accesible at www.taste-for-life.org.
Leer, J. and W. Karen. (2015). *Mod en smagspædagogik*, Skriftserie om smag #02, Aarhus Universitet: Institut for Uddannelse og Pædagogik.
Lynch, M. (2010). "Food Neophobia and Children's Vegetable Consumption. The Roles of Availability and Accessibility, Positive Pressure and Negative Pressure to Eat." Flinders University: School of Psychology.
Mann, A., et al. (2011). "Mixing Methods, Tasting Fingers. Notes on an Ethnographic Experiment." *HAU: Journal of Ethnographic Theory* 1(1): 221–243.
Markvorsen, S. (2015). "Smag, detektivarbejde og familiemysterier." In, S. Højlund, ed., *Smagsmysteriet. Smagen under lup.* Smag for Livet: Smag #03.
Menell, S. (1996). *All Manners of Food: Eating and Taste in England and France from the Middle Ages to the Present*. Illini Books edition.
Naccarato, P. and K. LeBesco (2012). *Culinary Capital*. London. New York: Berg.
Nielsen, M. (2012). "De fleste har en sukkerpolitik." *Fagbladet FOA* (1): 18–22.
Petersen, A.S. (2015). *Smag dig Frem: Masseeksperiment 2015: Resultater*. København: Dansk Naturvidenskabsfestival.
Pink, S. (2007). "Walking with Video." *Visual Studies* 22(3).
Science Daily. (2013). Institute of Food Technologists (IFT). "Taste Rules for Kids and Healthy Food Choices." *Science Daily*, July 16, 2013.
Sennet, R. (2013). *Together. The Rituals, Pleasures and Politics of Cooperation*. London: Penguin Books.
Spence, C. (2015). "Eating with our Ears: Assessing the Importance of the Sounds of Consumption on our Perception and Enjoyment of Multisensory Flavour Experiences." *Flavour.*
Stoller, P. (1989). *The Taste of Ethnographic Things. The Senses in Anthropology*. Philadelphia: University of Pennsylvania Press.
Teil, G. and A. Hennion. (2004). "Discovering Quality or Performing Taste? A Sociology of the Amateur, in M. Harvey, A. McMeekin and A. Warde, eds., *Qualities of Food*. Manchester: Manchester University Press.
Torralba, J. and B.A. Guidalli. (2015). "Examining Children's Food Heritage in Spanish Schools: A Process of Learning to Become 'Eaters-in-Context'?" *Anthropology of Food* (online) Patrimoines alimentaires enfantis 9.
Tuomela, R. (2007). *The Philosophy of Sociality. The Shared Point of View*. Oxford: Oxford University Press.
Ulloa, A.M., J. Roca and H. Vilaseca. (2017). "From Sensory Capacities to Sensible Skills: Experimenting with El Celler de Can Roca." *Gastronomica*.
Van Esterik, P. and R.A. O'Connor. (2017). *The Dance of Nurture: Negotiating Infant Feeding*. New York and Oxford: Berghahn Books.

9

Making the Multi-Dimensional Taste of Japanese Cuisine Public

Greg de St. Maurice

Introduction

"Where on your body do you register taste?" Chef Sonobe Shingo asks the elementary school students he teaches about Japanese cuisine. "Tongue!" students shout first. "Where else?" he asks. Usually students answer "Nose!" next. He continues: "Where else?" Eyes, ears, mouth, throat, brain, heart ... When the students have run out of suggestions, Chef Sonobe explains that the tongue can perceive sweet, sour, bitter, salty, and umami via the sense of taste. The sense of touch makes other foods astringent or spicy, he tells them. Then he asks what apples taste like. "Sweet!," "Sour!," "Sweet and sour!" they respond quickly. His next question is: "How about shiitake mushrooms?" Silence ensues.

Sonobe explains: "I get students to understand that it's not just apples and shiitake; the taste of foods can't be explained just in terms of tastes you perceive with your tongue" (Sonobe 2017: 188, my translation). When Chef Sonobe and his colleagues teach not only young Japanese people but also foreign chefs about the taste of Japanese cuisine they teach them to be attentive to all their senses and sociocultural factors, including ethical values, food attributes, and aesthetics.

In this chapter, I focus on the Japanese Culinary Academy (JCA) and its direct interaction with foreign chefs, publications aimed at professional cooks of Japanese cuisine abroad, and food education in local schools (from elementary through university). To find out about these activities, I interviewed members of the non-profit organization, including chefs and researchers. I also interviewed foreign cooks brought to Japan on special programs spearheaded by JCA members. Another source of data for this chapter is its series of professional cookbooks. I conducted the bulk of this fieldwork from summer of 2016 through summer 2017.

The JCA's approach to domestic food education and international outreach reminds us that social and cultural factors form tastes and mediate taste experiences, perhaps even more so for cuisine than for food. Through my ethnographic fieldwork, I saw that JCA members use culinary discourse and opportunities to cook and taste high-quality Japanese cooking first-hand to attune the senses and sensibilities of new audiences to the array of tastes and culinary philosophy that characterize Japanese cuisine. By making the "true" taste of Japanese cuisine public, they hope to rectify the popular reduction of Japanese cuisine to a limited repertoire of dishes (sushi, tempura, ramen, tonkatsu, etc.), ingredients (miso, soy sauce, wasabi), tastes (umami), and flavors (*yuzu*, matcha). To these chefs, sensory attributes and sociocultural frames are just as integral—if not more so—to the true taste of Japanese cuisine.

My investigation in this chapter of how Japanese chefs are making the taste of traditional Japanese cuisine public could not have taken place forty years ago. It is only in the past few decades that Japanese chefs have endeavored to take what was once closely guarded and often tacit, unverbalized professional culinary knowledge and actively share it not just with fellow chefs (potential competitors) but with cooks across the world and younger generations within Japan.

The impetus for opening up of the world of traditional Japanese cuisine was a perceived two-part threat: a weakening of culinary heritage at home and the proliferation abroad of what to chefs is a poor imitation of "authentic" Japanese cuisine. Making the taste of traditional Japanese cuisine public is meant to secure a measure of continuity domestically and ensure that even as Japanese cuisine evolves with its expansion across the globe, the discourse of "authenticity" and Japanese cuisine is informed by Japanese culinary philosophy and professional cooking standards. This is a very personal mission for JCA chefs, many of whom belong to families that have owned and operated a restaurant for generations. They have undergone years of grueling training in a traditional Japanese kitchen and wish to ascertain a sustainable future for this culinary heritage as they pass it on to the next generation.

The Complex World of Taste: Umami and the Taste of Japanese Cuisine

Taste, a term we use to mean so many different things, is hard to pin down. While people use the terms "taste" and "flavor" interchangeably in everyday English language conversation, for psychologists, neurologists, and biochemists, the term "flavor" conventionally refers to the combination of what is perceived by the senses of taste and smell. For them, a "basic taste" is one that cannot be created by combining any of the other basic tastes, can be discerned by unique taste receptors, and is commonly found in foods across the world. Natural scientists purposefully exclude sensory characteristics such as those perceived by retronasal olfaction (smells like cinnamon or smokiness) and somatosensory sensations (the coolness of menthol or the heat of chilies) from their scientific definitions of what they call "true taste," but they acknowledge that "somatosensory modalities such as texture and visual cues such as color also significantly influence the 'taste' of foods" (Chaudhari et al. 2009: 286) According to experimental psychologist Charles Spence, "what can be agreed upon is that there is an increasingly large body of scientific evidence demonstrating that the senses (*all* of them) interact in the multisensory perception of flavor" (2017: 33, emphasis in original). Our sense of sight determines whether we deem food appealing or not before we can even smell its aroma, much less take a bite. Texture and mouthfeel also matter, with attributes like crunchy, crispy, smooth, and rich often more desirable than stale, soggy, clumpy, and thin. Research has also demonstrated that sounds, music or background noise affects the taste of chocolate and toffee, for example, and can even inhibit one's ability to perceive specific tastes (Spence 2017: 30). The English language fails us here, as it lacks the vocabulary for talking clearly about multisensory experiences with food that go beyond tastes and flavors in terms of their conventional scientific definitions. Consequently, though I do in this chapter discuss the five basic tastes (sweet, sour, salty, bitter, and especially umami) when appropriate, I use the word "taste" as a verb and noun in ways that acknowledge that our entire sensorium is engaged when we eat and drink.

But to speak of the sensorium is to speak of a realm that has inextricable sociocultural dimensions. For one thing, tastes—such as those for sweet things, alcohol, or particular combinations of spices, for example—are linked to cultural categories and social prescriptions. Identities and affiliations exert their influence simultaneously, sometimes with what seem like contradictory results. Due to their ethnic identity, male Yadavs in Mathura, India, for instance, are supposed to practice vegetarianism, but as democratic, muscular, and secular politicians they are expected to develop a taste for chicken, goat, and whiskey (Michelutti 2010). Such examples remind us that consumption and connoisseurship, even when enacted in private, become practice in the social dimensions of taste (Douglas and Isherwood 1996 [1979]; Veblen 1994 [1899]).

Separating culture and "good taste" from the five basic tastes is not simple. In 1904, Charles Myers compared the terms different societies regularly used to describe the taste of solutions of sugar, salt, weak acid, and quinine (as sweet, salty, sour, and bitter essences). Myers gathered data on groups like the residents of the Torres Straits who use a phrase that literally means "tasting good" to describe both sweetness and saltiness (1904: 119), who do not have a linguistic equivalent for the term bitterness, and who "confuse" the taste-names for saltiness and sourness. Myers concluded that the people of the Torres Straits and other such "primitive peoples" had not evolved "sense-vocabularies" sophisticated enough to describe what he considered to be the four basic tastes. Several years after Myers published his paper, however, Japanese scientist Ikeda Kikunae published research declaring that he had discovered evidence of a fifth taste, that was prevalent in Japanese cuisine (Ikeda 1909). His research showed that where glutamic acid was present, this taste was perceived.

Ikeda's findings did not gain ground outside of Japan quickly, no doubt due to the fact that the particulars of his research were published in Japanese and also due to the difficulty of translating what seemed to be a culturally restricted "taste." Ikeda called this taste "umami," which can also mean "good taste" in Japanese. What would Myers have made of this?

It was Japanese researchers Kodama Shintaro (in the early twentieth century) and Kuninaka Akira (in the 1960s) who furthered Ikeda's research, finding that inosinic acid (in dry-smoked bonito) and guanylic acid (in shiitake mushrooms) are other umami substances, and that they have a synergistic effect on the taste of umami when combined with glutamatic acid (Yamaguchi and Ninomiya 2000). Yamaguchi and Ninomiya write that "[o]ne of the impediments to the recognition of umami as a basic taste may have been the lack of traditional words to describe it in Western languages," with English language translations including evocative but ambiguous terms like "amplitude," "mouth fullness," and "bloom" (Yamaguchi and Ninomiya 2000: 923S). Only after scientists in the United States found evidence for a receptor for umami (Chaudhari et al. 1996) did umami finally start gaining widespread acceptance as a basic taste.

Chaudhari and her team describe umami as "the meaty, mouth-filling, rich taste in many types of seafood, seaweed, fish, meats, and mushrooms" (Chaudhari 2009: 738S). Foods like Parmesan cheese, ripe tomatoes, and Jinhua ham contain umami—and West Africa's fermented locust beans possess a tremendous amount of free glutamic acid at 1700mg per 100g—but Japanese cuisine is said to be unlike any other cuisine because it is based around the taste of umami.

The Japanese seasoning agents miso and *shoyu* (soy sauce) add umami to dishes. Texts about umami and Japanese cuisine, however, place even greater emphasis on *dashi* soup stock. A staple for cooking Japanese food at home, in restaurants, and even in temple kitchens, the most typical dashi is made from *kombu* kelp (rich in glutamic acid) and dry-smoked bonito (*katsuobushi*, rich in inosinic acid), with regional and vegetarian variations. Most households today no longer prepare dashi from scratch, but rather purchase powdered dashi or pre-mixed bags containing kombu and katsuobushi that can be used to quickly prepare stock.

JCA chefs have been teaching new audiences to discern the "true" taste of Japanese cuisine beyond umami. Japanese attempts at defining Japanese cuisine—for instance in the application to have it registered as UNESCO intangible cultural heritage—focus not on dishes but aesthetics and principles, including seasonality, visual presentation, and quality standards for ingredients and flavors. These function as a frame that mediates the experience of tasting.

The Japanese Culinary Academy and its Outreach

Chefs, cooking school teachers, and other actors founded the Kyoto-based non-profit organization the Japanese Culinary Academy in 2004. Though Japanese cuisine had gained popularity and could be found across the globe by the early 2000s, chefs in Kyoto were concerned that what was being offered as "Japanese cuisine" and particularly high-end Japanese cuisine consisted of foods that mimicked Japanese food in certain respects but misrepresented Japanese tastes, aesthetics, and culinary techniques, and worse sometimes lacked the knowhow and sanitation standards that are important when serving raw fish, uncooked eggs, and other ingredients that must be carefully prepared. They saw Japanese cuisine as little understood overseas and often reduced to certain forms (the sushi roll, tempura, etc.) or flavors (umami, yuzu, matcha) in ways that glossed over the diversity and sophistication they understand to be part of Japanese culinary heritage (see Yomiuri 2004). Umami's success outside of Japan has raised the profile of Japanese food culture. But JCA chefs—and other actors in Japan—wish to go beyond what they feel is a superficial representation of Japanese cooking.

Priscilla Parkhurst Ferguson has argued that "The social survival of food in any given form depends entirely upon the critical discourse that translates the cultural presuppositions about food for the reader–diner ... culinary discourse secures the transitory experience of taste. It figures the material as intellectual, imaginative, symbolic, aesthetic" (Ferguson 2004: 19). Culinary discourse can transform "mere" food and "mere" flavors into cuisine for those who cook and those who eat, amplifying the cultural and social dimensions of taste experiences.

From its inception, the JCA perceived foreign chefs to be the natural and ideal audience for disseminating knowledge of Japanese culinary heritage abroad. In 2005 The Daily Yomiuri reported on the Academy's activities, stating that "the academy wants foreign chefs to learn the essence of Japanese cuisine, which originates in dashi soup stock, as well as the aesthetics of Japanese dishes" (Daily Yomiuri 2005). This notion of an "essence" is critical because it captures the sense that there exists some kind of core that can be experienced and understood (if not always taught or explained), even if it is constantly evolving. Again, this core consists of more than specific dishes, forms, or flavors: its components are both sensorial and sociocultural.

The JCA also coordinates locally oriented outreach in schools. A number of Kyoto's chefs had been visiting local schools to teach about Japanese culinary heritage independently, but once Japan passed a law establishing food education nationally in 2005, the JCA coordinated with the Kyoto City Board of Education to create what became the "Learning Food Education through Japanese Cuisine Curriculum." That year they taught sample lessons using the curriculum in five public elementary schools. In the years that followed, they visited an increasing number of schools and by the time the curriculum was officially put in practice in 2011, they were teaching in fourteen schools. The range of schools chefs visit today extends from elementary through to university. Demand has now reached the point where knowledgeable individuals who are neither chefs nor teachers—including homemakers with expertise in cooking, for example—can be certified to teach these lessons in chefs' stead.

Whether in elementary school or university classrooms, chefs aim to familiarize younger generations with the taste of plain dashi made from scratch and low-fat and low-calorie umami-rich foods. They see this as necessary because the vast majority of households today do not make their own dashi and some do not even regularly consume "traditional" Japanese cuisine (*washoku*). The result is that as children eat meals that include greater amounts of meat, fat, oils, and processed foods (if not salt or sugar) than in the past, their taste preferences become aligned with them, and as a generation they become "distanced from washoku" (*washoku banare*) and the "traditional" taste of Japanese cuisine. The extent of this is evident in a recent national campaign (in which a few JCA members are active) to designate November 24 as Washoku Day, urging elementary schools to offer students a traditional Japanese lunch and teach students about Japanese culinary heritage on this one day every year.

The Academy has increased its efforts to share the taste of Japanese cuisine with foreign chefs in recent years. What was once closely guarded knowledge is widely disseminated and recipes published in multiple languages. The first volume of *The Japanese Culinary Academy's Complete Japanese Cuisine, Introduction to Japanese Cuisine: Nature, Culture, and History* was published in English and Italian (and subsequently Japanese) in time for the 2015 World Expo in Milan (where a number of the chefs in the Academy cooked for the Japanese pavilion). The first of the volumes that followed was dedicated to seasonings and flavorings, the second to knife skills used for fish. Subsequent volumes are planned.

Up and coming chefs from thousands of miles away are today invited to learn to cook in Kyoto's most renowned kitchens. In the Academy's early years, foreign chefs attended cooking workshops in Japan and visited elite restaurant kitchens. In 2014, chef-owners of traditional Japanese restaurants successfully realized their goal of establishing a two-year work visa for foreign chefs seeking to learn to cook authentic Japanese cuisine. A handful of chefs have taken advantage of this visa since the first arrived in early 2014 to work in Murata's restaurant, Kikunoi. Due to the JCA's efforts, Kyoto has become the only city in Japan with such a work visa system in place. At the time of writing this chapter, seven chefs have obtained this visa, two have graduated, and it has been extended to allow chefs to work for a restaurant for up to five years.

Kyoto institutions have joined restaurants in other parts of the country to create another program for foreign chefs to learn Japanese cuisine in Japan. In 2016–2017 this program, sponsored by the Ministry of Agriculture, Forestry and Fisheries and organized by the Japanese Cuisine and Food Culture Human Resource Development Committee, offered fifteen foreign cooks a six-month apprenticeship after an intensive language and culture course and training in Japanese cuisine at Le Cordon Bleu Japan in Tokyo. A second incarnation of this program is being implemented for 2017–2018.

Old Words and New Words: Bridging Experiential Difference

Culinary discourse can, in Ferguson's words "secure the transitory experience of taste" and bring about the "social survival of food." Language matters all the more when it comes to cuisine rather than food. Sidney Mintz argued that an authentic cuisine (as opposed to restaurant cuisine or a national cuisine, in his mind) only emerged where one had a community that cared about the styles of food they ate and made this a matter of discussion (Mintz 1996). Moreover, cuisine is a form of communication (Farrer 2015: 4), with reception an integral but hardly straightforward component.

The sharing of taste, especially across cultures and languages is complicated by the ways that language, cultural frames, and taste experiences are tied together. Because the taste of umami is considered to be an essential aspect of Japanese cuisine, it serves as a good example of how the JCA has used a combination of old and new words to articulate the taste of traditional Japanese cuisine and attract new audiences. The JCA (along with other actors including the Tokyo-based Umami Information Center) introduced umami to foreign chefs as a key concept for grasping Japanese cuisine and its aesthetic ideals. Indeed, umami appears on the first pages of the JCA's *Introduction to Japanese Cuisine*, where historian Kumakura Isao states, "The basic flavor of Japanese food is umami" (Kumakura 2015: 9). Umami appears throughout the text, most frequently when discussing dashi and the "essentials" of Japanese cuisine as well as fermented seasonings. Following an explanation of why it is a basic taste, the text shows that umami is neither a taste restricted to Japanese cuisine nor one that only

Japanese people can discern: "Often described as 'savory.' umami is associated with foodstuffs as diverse as kombu kelp, aged cheese, and ripe tomatoes" (104). Later, this use of examples is expanded upon: "Many kinds of soup are made around the world by boiling meat, bones, and vegetables, and almost all of them are built around the taste of umami" (106). In the second volume of the series: "The umami taste may have been identified in Japan, but umami-rich broths very close to Japanese dashi, like Russian borscht soup made from beets and beef, or high-quality Chinese tang broth made with Jinhua ham, have been used in cooking since long ago in places all over the world" (7).

As with umami, in explaining flavor aesthetics, Japanese terms are used. While it has yet to be used in the book series, Academy chefs often emphasize the importance of the concept of *hin'i* (alternately, *hin*), a subdued and refined taste that is created with precision and restraint (de St. Maurice 2012). There is a parallel here with French cuisine, as the use of French cooking terminology contributed to the Frenchness of the cuisine as it was taught and promoted (Ferguson 2004). Words that French chefs of modern French cuisine purposefully invented "for the tastes meant to seduce the diner"—like *ragout* and *garniture* (garnish)—have had staying power (Peterson 1994). New words such as these can establish a cuisine as distinct from others in terms of sensory experience and culinary philosophy. *The Japanese Culinary Academy's Complete Japanese Cuisine* is replete with Japanese words accompanied by English translations (e.g. *suppon* soft-shelled turtle, *asatsuki* chives, *mochigome* glutinous rice) and others like *bekko-an* sauce or *kobujime* curing that are followed by explanations and instructions.

To help Japanese taste aesthetics and terminology take root overseas, programs bringing promising chefs from overseas to train in Kyoto kitchens are similarly intended to teach a comprehensive vision of taste beyond dishes and umami. The work visa established by the JCA and Kyoto City, however, is only several years old and kitchen staff are still in the process of adjusting to working with apprentice chefs who lack familiarity with the Japanese language and taken-for-granted ways of doing things in Japan. A further complication is that Japanese kitchens have not traditionally been workplaces where techniques and recipes were taught outright. One frustration I heard voiced in interviews with Japanese and non-Japanese chefs alike was that formal training or learning opportunities in traditional kitchens were rare and that it was the responsibility of newcomers to find ways to get senior colleagues to teach them. Making the taste of traditional Japanese cuisine public may have been a conscious choice but the process has not been entirely smooth. It might be accurate to say that Japanese chefs are themselves formulating the vocabulary and the tools to create the tutelage or frame that can build the subjecthood requisite for the preparation of "authentic" Japanese cuisine not just across languages and cultures, but also in a condensed period of time.

Visual Elements

Kaiseki cuisine and fine Japanese dining in general place considerable importance on presentation, and the taste the JCA's members make public includes visual aesthetics.

There are obvious visual means of doing so. The JCA uses books, website pages and posts, and other paper and electronic print media to present visual explanations and representations to foreign chefs. *The Japanese Culinary Academy's Complete Japanese Cuisine* is replete with gorgeous high-quality photographs that include top-shelf fresh fish and vegetables, production methods for fermented foods, and specially prepared dishes served on carefully selected tableware that is itself a work of art. The volume's stunning photography is not merely a means of attracting buyers or even impressing audiences. It is also meant to be educational, with the books presenting visual aesthetics believed essential to "true" Japanese cuisine.

For instance, in *Introduction to Japanese Cuisine* a section titled "Artistic Awareness" introduces foreign chefs to Japanese aesthetic concepts like *wabi* and *sabi*, concepts that together reflect a beauty found in simple things that have been worn down by time. On the page opposite the explanation of *wabi* and *sabi* is a photograph of the dish "Grilled Barracuda with Miso *Yuan-yaki* Baste." The glossy earthy and metallic colors of the pieces of skin-side up barracuda are echoed in the rustic dish on which they have been arranged.

Directions for the appropriate plating of dishes are part of the recipes in *The Complete Japanese Cuisine*. An important component of plating that is evident in the photographs but not explicitly spelled out is the selection of tableware. This characteristic of Japanese cuisine can be traced back to the influence of artist and restaurateur Kitaōji Rosanjin in the early twentieth century (see Stalker, forthcoming). Dissatisfied with the quality of the dishes and serving vessels he could afford for his restaurant, Rosanjin (known by his first name) went so far as to make his own, breaking from tradition by borrowing pottery techniques from all over Japan rather than perfecting one particular style. When Rosanjin first visited the United States he was pleasantly surprised by the meals he ate, but declared that food presentation, tableware included, left much to be desired. Those JCA affiliated restaurants with the longest history and financial resources have the famous potter's wares in their collection of tableware and collaborated with the Museum of Modern Art Kyoto for a 2015 exhibition titled "Kitaoji Rosanjin: A Revolutionary in the Art of Japanese Cuisine," a kind of educational outreach aimed at the general public.

The notion that presentation matters to the taste of Japanese food is not restricted to haute cuisine. In fact, teaching plating/food arrangement is an official goal of the JCA's food education lessons. Chef Inoue Katsuhiro of the restaurant Imasa teaches elementary school students to pay attention to the visual appeal they can achieve via color and the three-dimensional arrangement of ingredients, for example. Students watch Chef Inoue demonstrate this and then have the opportunity to create—and later consume—dishes whose visual appeal contributes to their deliciousness. For his part, Chef Sonobe asks students to consider how a *dashi maki tamago* (Japanese-style omelet made with dashi) cut into clean, uniform slices will taste different from one that is cut sloppily and haphazardly. The students do not make a *dashi maki* in his class, but the small groups are reminded that the same tenet applies when they peel daikon, cut *mizuna* (a Brassica variety that is considered a Kyoto vegetable), and arrange the final product in bowls for each student in their class.

Experiences Preparing and Tasting

As important as words and images are in communicating taste, chefs in the JCA have discovered their limitations. Chef Murata puts it this way: "How do you explain foie gras or truffles to someone who's never had them? You might be able to read the sheet music for Beethoven, but unless you've actually heard his music, you won't know what it sounds like." Ninomiya Kumiko of the Umami Information Center, who collaborates with the JCA, echoed this sentiment: "If I just explain that umami is the taste of glutamate, can you imagine it?" Structured and guided opportunities for people to familiarize themselves first-hand with Japanese cuisine accomplish what words cannot.

This is especially important because the taste of Kyoto cuisine is not one that the uninitiated can easily discern and appreciate. The JCA's early efforts, aimed at elite foreign chefs who were invited to Kyoto to experience Japanese cuisine behind the scenes, were not resounding success stories. Language and cultural barriers and time constraints imposed limits on the scope of what the invited chefs could learn.

To engage in deeper outreach abroad, the JCA then decided to create immersive, long-term opportunities for young foreign chefs to come to Kyoto to learn how to create traditional haute cuisine. The cooks participating in these programs, like their Japanese peers, are trained to be hyper aware of their surroundings and to take note of how their superiors are accomplishing their tasks. Interviewees referred to this approach as "watching and stealing" (*mite, nusume*) knowledge. It is not just sight that is involved in kitchen learning and the acquisition of taste discernment. For instance, one has to be able to determine that a meat has been properly cooked through one's sense of touch, either by applying a skewer that has been inserted inside the meat to the area underneath ones lips or by gently squeezing the meat with one's fingers. The temperature and resistance of the meat indicate the extent to which it has been cooked. Cooks are trained to taste with their entire sensorium and their skills and knowledge are explicitly talked about as embodied.

First-hand experiences with the senses are also a key component of food education aimed at young students in Kyoto. I witnessed this when I attended the dashi tasting event at Kyoto University in December of 2016. After Chef Sonobe's lecture on kombu, katsuobushi, and dashi, the students divided into groups assigned to booths run by the five restaurants in attendance. The five chefs ladled their dashi into tiny plastic cups for students to sample and compare. Of course, it is neither customary to consume dashi on its own, nor to drink it out of a disposable plastic cup. But every student who attends is also offered *suimono*, featuring a particular restaurant's dashi stock with *yuba* (tofu skin), mizuna, and yuzu (Japanese citrus variety) peel in a fine lacquer bowl. Each restaurant prepares this simple dish in its own way, varying preparation techniques and amounts. Chef Tanaka Nobuyuki of Tsuruse demonstrated how students should take the warm bowl into their hands, hold it under their noses, slowly open the lid, inhale the fragrant steam, appreciate the visual arrangement, take a sip of liquid, use their chopsticks to grasp the solid ingredients, and then take their first bite, mindful of the array of textures, flavors, and other sensations. This was a carefully structured and guided taste experience, preparing students to engage their senses and knowledge and understand dashi when they sipped and ate their bowl of suimono.

Values

As the previous section hints at, in teaching *how* to taste, JCA chefs are also teaching tasters *what* they should be sensitive to. Perhaps surprisingly, values and notions of heritage factor into this. At its inception the food education curriculum was modeled around Jacques Puisais' methods for teaching French children about food and gastronomy. However, this approach did not catch on, Chef Sonobe, who leads the food education project for the JCA, told me. Sonobe elaborated, "Because the Puisais method is especially about the sense of taste and Japanese cuisine is not just about the sense of taste, we incorporated ingredients and cooking, energy necessary for life and feelings of gratitude, and also consideration for others into the contents [of the curriculum]." While chefs could—and did—concentrate on the sense of taste or even flavors, this felt disingenuous and beside the point to them.

After all, the JCA does not simply want to teach the taste of traditional Japanese cuisine for its own sake, but because they hope to extend its history. "We want food to become a topic of conversation, to generate communication. We want students to go home and tell their parents about the dashi they made at school," Sonobe explained. There seems to be a sense of urgency in teaching values domestically to alleviate "washoku banare" (distancing from washoku) and make the taste of traditional Japanese cuisine an intimately familiar one. For this reason, the food education curriculum takes as its basic unit not the individual child, but the household.

Chef Sonobe and Professor Matoba Teruyoshi made it clear that the food education curriculum is built around a core set of values. Each point being taught is paired with a set of ethical goals related to cooking and eating. Honing students' ability to taste with the five senses, for instance, can help develop their *ikiru chikara* (zest for life, ability to live) by teaching them how to distinguish between good and bad, safe and unsafe foods. Teaching about gratitude for delicious food, meanwhile, is meant to develop *kansha suru chikara* (ability to be thankful). And chefs teach students to put their hearts into cooking in the aim of helping them develop *omoiyari no kokoro* (considerate hearts, thoughtfulness).

While these values are sometimes emphasized in outreach, they are taught more indirectly. Take the lesson of thankfulness for delicious food, for example. In May 2017 I watched Chef Nakahigashi Hisao teach fifth grade students about dashi and how to cook it. Chef Nakahigashi's lecture followed the same general pattern his colleagues use: he explains how bonito is caught, transported, dry-smoked and mold-cured, and the time and effort that go into the harvest, drying, and aging process for kombu (and how all conditions have to be optimal to create the highest grade kombu). Dashi is taught as a uniquely Japanese phenomenon and superlatives are often used to accentuate this: dashi is "the quickest" soup stock to prepare (though a great deal of time and labor are invested in the production of its basic ingredients) and a katsuobushi fillet (before it is shaved into flakes to make dashi) is "the hardest foodstuff on earth."

The focus of this lesson is not that dashi is delicious because it is full of umami, but that it is a delicious heritage food that deserves respect and gratitude.

But Chef Nakahigashi made this lesson even more personal for students. At Chef Nakahigashi's request the students had grown daikon in their schoolyard specifically for that day's lesson. Nakahigashi reminded the students of the three months they had tended the daikon before harvesting them, drawing a parallel with the time and human labor involved in kombu harvesting and katsuobushi production. This, students are reminded, is why before eating Japanese people customarily give thanks using the expression, "Itadakimasu."

Efforts are made to share values and notions of culinary heritage with foreign chefs, too. Chef Murata explains that the first volume of *The Complete Japanese Cuisine* series includes material about "the Japanese natural environment and landscape, history, culture, and basic culinary techniques . . . to provide a unified understanding of and comprehensive introduction to the background and context within which the arts, techniques, and wisdom of Japanese cuisine have developed. True appreciation of a cuisine begins by gaining an understanding of such a background" (2015: 7).

Kumakura's short essay in the introductory volume notes that "Japanese people are attuned to nature and keenly aware of their reliance on its bounty. They express gratitude for the blessings of nature with the customary expressions "Itadakimasu"—I gratefully receive the blessings of this food—before eating and "Gochisosama"—I have partaken of the feast—after eating" (Kumakura 2015: 9). The essay goes on to link this to the Japanese appreciation of seasonality. Kumakura includes this material not because he expects foreign chefs to adopt Japanese values and customs, but rather because—like Chef Murata—he wants chefs to appreciate them and how they have influenced Japanese cuisine.

Conclusion

To extend the social life of traditional Japanese cuisine, the Japanese Culinary Academy has begun to teach new audiences about what authentic Japanese cuisine should taste like and how to make it. In their efforts to reinvigorate domestic culinary heritage and prevent Japanese cuisine from being broadly misrepresented abroad, they have paired lessons about Japanese cuisine with opportunities for new audiences to experience it first-hand, through cooking and eating.

Representing the taste of Japanese cuisine with a dish or even a meal is a faulty proposition. The taste that the JCA is making public lies not in a set of flavors, dishes, or techniques. It is made by informed cooks and perceived by awakened eaters and thus produced today in the very act of being made public. Another way of looking at it is that JCA chefs understand that any taste experience is filtered by our background, identity, and the information we have about the food being tasted. They wish to improve (from their perspective) the filter through their outreach, encouraging foreign cooks to turn to Japan for notions of "authentic" Japanese cuisine and instilling in younger generations of Japanese people an intimate connection with what is considered to be their culinary heritage. For chefs in the Japanese Culinary

Academy, all of this—sensory attributes, values, ethics, and aesthetics included—is a matter of taste.

References

Chaudhari, N., E. Pereira, and S.D. Roper. (2009). "Taste Receptors for Umami: The Case for Multiple Receptors." *The American Journal of Clinical Nutrition* 90, supplement: 738s–742s.

Chaudhari, N., Hui Yang, C. Lamp, E. Delay, C. Cartford, T. Than, and S. Roper. (1996). *Journal of Neuroscience* 16(2): 2817–3826.

Daily Yomiuri. (2005). "Refining Tastes/Japanese, French Chefs Share Culinary Ideas." November 7, 2005.

de St. Maurice, G. (2012). "Savoring Kyoto: Sensory Fieldwork on the Taste of Place." *Etnofoor* 24(2): 110–22.

de St. Maurice, G. (2017). "Kyoto Cusine Gone Global." *Gastronomica* 17, no. 3 (2017): 36–48.

de St. Maurice, G. (Forthcoming). "Sensing Bodies at the Center in Today's Traditional Japanese Restaurant Kitchens." *Japanese Studies*.

Douglas, M., and B.C. Isherwood. (1996 [1979]). *The World of Goods: Towards an Anthropology of Consumption*. London: New York: Routledge.

Farrer, J. (2015). "Introduction: Traveling Cuisines in and out of Asia: Toward a Framework for Studying Culinary Globalization," in *The Globalization of Asian Cuisines: Transnational Networks and Culinary Contact Zones*, 1–19. New York: Palgrave Macmillan.

Ferguson, P. Parkhurst. (2004). *Accounting for Taste : The Triumph of French Cuisine*. Chicago: The University of Chicago Press.

Ikeda, K. (1909). "Shin chōmiryō ni tsuite." *Tokyo kagaku kaishi* 30(8): 820–836.

Imai, S. (2015) "Umami Abroad: Taste, Authenticity, and the Global," in *The Globalization of Asian Cuisines: Transnational Networks and Culinary Contact Zones*, 57–78. New York: Palgrave Macmillan.

Ishige, N. (2012). *L'art Culinaire Au Japon*, translated by E. Marès. Paris: Lucie editions.

Japanese Culinary Academy. (2017). *Flavor and Seasonings: Dashi, Umami, and Fermented Foods*. Tokyo: Shuhari Initiative.

Japanese Culinary Academy. (2015). *Introduction to Japanese Cuisine: Nature, History and Culture*. Tokyo: Shuhari Iniative.

Kumakura, I. (2015). "What Is Japanese Cuisine?" in *Introduction to Japanese Cuisine: Nature, History and Culture*, 8–9. Tokyo: Shuhari Initiative.

Michelutti, L. (2010). "We are Kshatriyas but We Behave like Vaishyas": Diet and Muscular Politics Among a Community of Yadavs in North India. *South Asia: Journal of South Asian Studies* 31(1): 76–95.

Mintz, Sidney (1996). "Tasting Food, Tasting Freedom: Excursions into Eating, Culture, and the Past." Boston: Beacon Press.

Myers, C.S. (1904). "The Taste-Names of Primitive Peoples." *British Journal of Psychology, 1904–1920* 1(2): 117–126.

Peterson, T. Sarah. (1994). *Acquired Taste: The French Origins of Modern Cooking*. Ithaca: Cornell University Press.

Rath, E. (2014). "New Meanings for Old Vegetables." *Food, Culture, and Society* 17, (2): 203–223.

Sonobe, S. (2017) "Ryōri nin no torikumi jirei." in *Dashi no kagaku*, ed., Teruyoshi Matoba and Naoto Tonouchi, 188–193. Tokyo: Asakura shōten.

Spence, C. (2017). "Multisensory Flavour Perception." In *The Taste Culture Reader: Experiencing Food and Drink*, ed., Carolyn Korsmeyer, 29–36. New York: Bloomsbury.

Stalker, N. (Forthcoming). "Rosanjin: The Roots of Gurume Nationalism." In *Devouring Japan*, ed., Nancy Stalker. Oxford: Oxford University Press.

Veblen, T. (1994 [1899]). *The Theory of the Leisure Class*. New York: Dover Publications,.

Yamaguchi, S. and K. Ninomiya. (2000). "The Use and Utility of Glutamates as Flavoring Agents in Food." *Journal of Nutrition* 130(4) supplement: 921S–926S.

Yomiuri Shimbun. (2004). "Nihon ryōri wo sekai ni hasshin ryotei shujinra 'akademi-' setsuritsu." August 26, 2004.

Yoshihara, A. (2005). "Refining Tastes; Japanese, French Chefs Share Culinary Ideas." *The Daily Yomiuri*, November 7, 2005.

10

Sharing and Transmitting Taste in a Professional Danish Restaurant Kitchen

Jens Sejer Østergaard Rasmussen

Introduction

Food taste may appear to be a very private experience of building a relationship between one's senses and an object, with the capacity to impact through aromas and tastes. It is only you who, when biting, taste the luscious, sour juices of an apple or the crispy, salty outside of a tender, fried chicken leg. But taste is something that is socially established, especially in the restaurant kitchen. In this chapter, I present an empirical study of taste in the restaurant kitchen to illuminate how it is also socially constructed, shared and experienced. Professional chefs constantly work to fit into specific ways of understanding taste by both mental and practical adherence with the collective know-how that is established among chefs in the restaurant kitchen. I argue that taste is what binds chefs to materials in specific ways but also to the social milieu in which they work, and taste is actively construed through social synchronizations of approaches to raw materials. This argument will be underscored by empirical data taken from fieldwork carried out in 2014 and 2015 in close collaboration with six chefs in the restaurant kitchen of the Danish restaurant Restaurant Mellemrum—a gem of a cosy gourmet restaurant, well situated in an alley in the city centre of Aarhus, the second largest city of Denmark.

Throughout the last decade, Aarhus has ridden the wave of culinary interest that has permeated Denmark, and most other northern and Scandinavian countries. Within the last ten years, the Central Denmark Region, where my field site was located, has been a focus of attention from national as well as international media, and from prominent chefs around the globe due to its thriving culinary landscape. In general, Denmark is characterized by an amplified focus on food through the growing attention given to healthy food, sustainable cooking and ingredients, and professional chefs by the mass media and in politics. As Jonatan Leer remarks, the interest in food among the general public in Denmark has given rise to an army of celebrity chefs that roam the Danish media (2013). The chefs not only cook on TV but also function as guides for the "good life" by focusing on health, ethics, "nature romances," hedonism and so on (Leer 2013: 43). During the last decades the natural habitat of the Danish chef has even seemed to be in a political context in TV. Nevertheless, the fact that the work of Danish chefs has been extended beyond the restaurant kitchen does not necessarily mean that they have left the restaurant kitchens. Rather these discourses illuminate how food-making in general can be seen as a continuum for negotiations of social meaning that expands beyond mere nutritional values. Professional chefs are intensely involved with food-making as a basic part of their everyday life and the fact that the chef has appeared on TV screens and on the political stage does not take away from the fact that they are particular experts in one thing, namely making tasty food for people.

The Danish food scene has attracted much attention given the New Nordic movement, which has spread throughout the world as a "return" to local cooking traditions and ingredients (Risvik et al. 2009: 7; cf. DeSoucey 2010; Trubek 2007). Most high-end modern Danish restaurants, such as the one discussed in this chapter, are in some way entangled in this approach to cooking due to their eagerness to satisfy their Danish customers, who also find satisfaction in eating local and traditional food (see Byrkjeflot et al. 2013). In many ways, the list is endless when it comes to the tendencies and trends that define the variegated landscape of food knowledge in Denmark, whether we speak of the healthiness, tastefulness, festiveness, sustainability, political or social statuses that are vocalized and accentuated around specific foods. All these come to be a bank of values that chefs draw from, when constructing their own food attitude. Therefore, chefs are "powerful cultural brokers," as they constantly juggle in an embrace or rejection of knowledge surrounding the food materials and cooking techniques that they use in restaurant kitchens (Beriss and Sutton 2007: 1–3). Because of their intense involvement in food making, I see professional chefs and their concerns, knowledge, values, practices and experiences in relation to food as a window into larger processes of establishing taste in a Danish context, and as I experienced during my fieldwork, chefs are experts in sharing these facets through food and language (cf. Fine 1996 and Fine 2009).

(Dis)regarding the Basics and Learning to Really Taste

All the chefs at Restaurant Mellemrum are educated at different cookery schools that all provide them with specific skills about cooking methods and access to standardized knowledge about cooking, eating, ingredients, and flavours. The first time I asked any of the chefs about their relationship to taste, they all referred to the five basic tastes: sweetness, sourness, bitterness, saltiness and umami. When I asked them where they had learned about these things they generally answered "at cookery school." There was no doubt that these five categories were considered as objectively existing, in a scientific manner. Simon, the youngest apprentice chef, told me that "it makes it easier to make balance between the different elements of a dish," and he continued, "you can use these tastes as a way of thinking about what a dish needs, and what it doesn't need." As Simon remarks, defining food objects by their objective impact on the taste buds with emphasis on one or more of these five taste categories, makes it easier to imagine and decipher different facets of a dish mentally. The chefs saw this knowledge about the physiology of taste as basic cookery knowledge that one must be able to account for in relation to every dish one serves. But this was more of a thought principle than a practical one, since the chefs rarely accounted for the dishes they made with reference to only these categories. Despite the absence of empirical accounts for the existence of these five tastes in dishes that were cooked in the kitchen, these categories seemed to offer the chefs an explanatory model for comprehending the relation between the mental experience of flavor and a sensory experience of ingredients. In many ways the chefs seemed to echo Priscilla P. Ferguson's claim that the conventional spectrum of physiological tastes is a productive model for "a working chef" to secure "his authority by calling upon the resources of science" (Ferguson 2011: 377). The taste spectrum is useful when trying to give authority to a dish and when trying to understand taste as something that happens with the taste buds, but as Ferguson remarks, rarely are foods just sweet, sour, salty, bitter and umami, but a combination. Moreover, foods gain tastes through links to identity, emotions, memories and other experiences (Ferguson 2011: 388).

Most of the scholarly knowledge that had provided the chefs with access to work in the restaurant kitchen was used to establish authority in a cooking environment. Scholarly knowledge provides chefs with the cultural or culinary capital to differentiate themselves from non-cooks on a formal level (Bourdieu 1984; see Naccarato and Lebesco 2012: 2). Their scholarly knowledge allows the chefs to position themselves in the field of status and in turn this status provides them with access to an important arena, namely the restaurant kitchen.

Among the chefs, the restaurant kitchen was perceived as the most important place for learning about taste through experience. As one of the apprentice chefs, Mads, explained to me: "I want to do something. That is how I learn. Of course we cook at the school, but that is not real cooking . . . like that in a restaurant kitchen." For Mads, the conventional school form, where a teacher teaches, and a bunch of students learn from it, did not necessarily represent how one learns to cook and taste. In the cookery school, apprentices often engage directly with ingredients and they do basic skill and hygiene training, but despite the fact that the chefs were being taught in kitchens at the school they generally agree upon the fact that the school arena was not the culinary "real world." The teacher's words did not amount to learning to cook; rather one learns to cook by cooking; and therefore, one learns to taste by tasting. Nevertheless, the explanatory models for taste with which the chefs have been equipped, are modes of verbalizing taste in a shared manner. In this sense, what has been taught about taste at cookery school is a platform for sharing impressions of food in a scientific manner, which serves the purpose of strengthening the social position of the cook as a chef. But practically speaking, taste has a hinge in the corporal engagement with ingredients, and in relation to chefs, the restaurant kitchen is the "real" or right context for this engagement with taste.

Rhubarb

What they *do* in the kitchen is central to chefs' perceptions of taste. Among my informants, there was a shared idea about the fact that, when a chef enters a kitchen for the first time, s/he needs to be taught how "things are done" there. This process of getting to know not only the kitchen and the people in there, but also the tastes, and the techniques that need to be used to produce these tastes, was of course especially visible in relation to the apprentice chefs, since they had less experience in the kitchen than the other chefs. Thoughtful and compelling research on the master-and-apprentice setting (cf. Argenti 2002; Herzfeld 2003; Hill and Plath 2006) has shown the complexity of these power-laden relations, so I will not be attempting to scrutinize this aspect of apprenticeships in this context. Rather I would like to question, what the transmission of techniques between people working within a kitchen—whether from master to apprentice or from chef to chef—tells us about taste, and more importantly, how taste is transmitted from one person to others, and back again?

Among chefs there is an idea about working "openly," which means that it should be possible for other chefs to observe what you are doing, how you are cutting, how much butter you put in the pan and so on and so forth. Chefs said that this openness was one of the most important ways of ensuring great taste in the kitchen, because it affords a continuous flow of knowledge about other chefs' techniques due to the possibility of perceiving what is going on in detail. To chefs, openness is based on both a willingness to share ideas, understandings and stories about cooking with the other chefs and a corporeal openness, where one stands in a position so that other chefs can observe how you handle ingredients and equipment. Openness is thus a crucible for exchange of techniques surrounding food. This openness does not have the overtone of monitoring the outcome of cooking practices, but it is an important part of an overall rule of sharing techniques and knowledge in the kitchen. Through openness, both cooking knowledge and taste are shared.

Marcel Mauss was one of the first anthropologists to underline the importance of scrutinizing how habits are embodied, and his reflections on "techniques of the body" are very relevant in relation to the transmission of taste among chefs (Navaez 2006: 60). Rafael Navaez remarks that Mauss considers the body as born in nature, but ultimately realized through culture. This brings forth an important aspect of techniques in relation to cooks. As Mauss remarks, by being born in nature, the body is installed with endless potentialities for usage, and through culture these potentialities are released (in Navaez 2006: 60). For Mauss, the body is an instrument for technology: "a technology that is traditional and works efficaciously within a certain social order" (ibid.). Thus, technologies of the body, for Mauss, are ways of comporting oneself in a social setting, and these comportments arise from the relation between knowledge and experience (traditions), and corporeality through cultural practices. That is to say, the body acts as a crucible of social meaning; you do things a certain way because they make sense not only in a practical respect but also in a larger social context. Thus, people adjust and share postures, movements and gestures in a process of comportment with the social milieu in which we live. Therefore, the way we carry our body becomes a matter of shared practices, and this is the point that will be drawn into analysis in this section. To understand the transmission of taste as a part of sharing bodily techniques, the ways that chefs handled rhubarb will be the vantage point for this exploration.

One morning in the kitchen at Restaurant Mellemrum, a large box of about ten kilos of rhubarb stems had arrived at the restaurant. They were supposed to be used for rhubarb compote, so Simon was asked to clean them and cut them into pieces before they could be put on to simmer. Cutting rhubarb seems like a simple task, but as will be shown in the unfolding of the processing of the rhubarb it can be done in several ways and these different ways purport different potentials of the rhubarb as used in dishes:

> Simon takes up a small vegetable knife and starts cutting the rhubarb across in large chunks (see C in Figure 10.1), then he looks at Mads who has also started cutting the rhubarb on the other side of the counter of the cold section of the kitchen. Simon looks for a few seconds and then he starts cutting the rhubarb in smaller pieces (see A in Figure 10.1). Then Mads looks at Simon's cutting techniques and he quickly grabs Simon's knife because "the rhubarb has to be cut more sharply," and the vegetable knife that Simon used was too small to make an even cut through the rhubarb stem. Mads takes a larger knife and shows Simon how to cut the rhubarb correctly (see B in Figure 10.1). As Mads shows Simon, Simon asks Mads why they had to be cut in that way since it ultimately has to go into the pot and simmer into a mash. From a pragmatic perspective, it is understandable why Simon questions the importance of cutting the rhubarb with such precision because it seems inefficient in respect to how much time is used on this task. But, as Mads explains, the cutting is not a mere matter of just cutting: ". . . this is the way we cut rhubarb for compote, cause cut this way and with this sharp angle, the size of the chunks are perfect and they will not be totally dissolved in the compote and they do not become too chewy and sour, which would be the case if they were cut into larger chunks because of the stringy nature of the stems." Then Simon takes the knife again and continues cutting the rhubarb by imitating Mads' cutting technique.

By looking at the cutting of rhubarb as a bodily technique it becomes clear how the sharing of techniques is a way of transmitting certain ways of perceiving and producing taste in the kitchen. One cut makes the ingredient appear and taste one way and another cut, another way. Thus, beyond the importance of thoughtful processes of recognition and reflection (as referred to in relation to the five taste categories), perceptions of taste also lie in the corporeal choices you make as a chef. The openness that is implied in kitchen practices allows for a transmission of knowledge about ingredients, and ways of handling them, which are in tune with the desired tastes. This transmission applies not only to simple techniques of cutting, but also extends to the making of complex sauces, mixing of dough and searing of meats. Experiences and explanations are shared between the chefs, which give meaning to different techniques

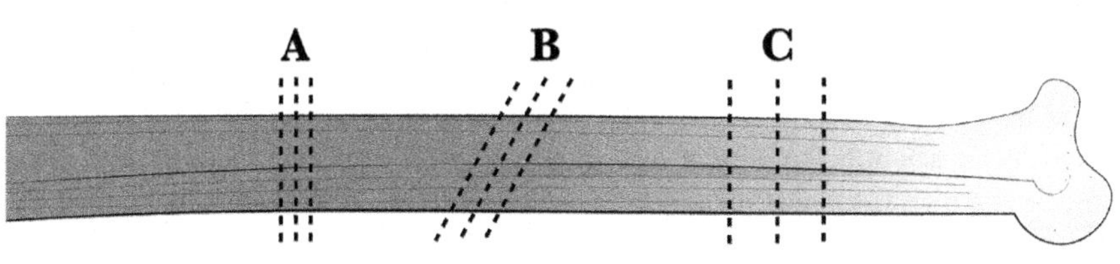

Figure 10.1 Different cuts shown on a rhubarb stem. Photo by Jens Østergaard Rasmussen.

and ingredients. These transmissions adjust the chefs' practices in a continuous striving for collective synchronization of knowledge and skills surrounding taste. Thus, through openness a framework of shared techniques of the body and taste practices is established.

A Turning Point

To underscore the social implications of establishments of shared frameworks for tastes in the kitchen it is important to underline that these adjustments are not only techniques of the body. They are also understandings and histories of taste together with the issue of work rhythms. As David Sutton has remarked, memories do impact the ways in which we perceive tastes (Sutton: 2010), and this aspect of taste is also important when considering the sharing of tastes in a restaurant kitchen. Ronni, one of the older chefs, refers to a specific occasion during his apprenticeship:

> I remember when I was a kid I was disgusted by the smell and taste of seafood, but I remember the turning point of this disgust. I remember an episode where I was on the sidelines, like most other apprentice chefs are when they are working with new dishes. But at one point the head chef asked me to pour cognac into the bowl and light it . . . to flambé it! I had seen him do it before, but never done it myself. Nevertheless, I took the bottle and poured in the cognac. I lit the pan and I still remember the sweet smell of the shells from the seafood as they simmered and gave off their beautiful aromas. Of course, I also remember how the head chef seemed to give an accepting nod as he walked off to leave me with the rest of the preparation of the dish. Maybe stupid of him? It feels like that episode made me understand the taste of the seafood and suddenly I felt like I fitted into this kitchen.

In his narrative, he remembers two relationships to seafood, which are disgust and appreciation of taste. Despite the fact that disgust has often been seen as a somewhat rigidly located feeling sticking to certain objects (see Sarah Ahmed 2004), Ronni explains how his interaction with the previously disgusting objects enabled him to transcend disgust by encountering it with an appealing aroma which he produced. Anthropologists of phenomenology of skill and embodied practices have often tried to go beyond the idea of a linear attainment of knowledge from a master to his apprentice. In his study of minaret building and apprenticeship in Yemen, Trevor H.J. Marchand explains how his participation in the apprenticeship gave him insights into how the acquisition of craft knowledge happened without explicit verbal orders coming from the master and without any systematically structured programs (2001). What Marchand points to is the fact that a crafted object does not hold a certain form by linguistic or scientific knowledge, but rather by interrelationship between the social milieu in which the object needs emplacement, and the skills of the one crafting the object. Marchand argues that the social interaction with the master, and his physical engagement in brick carving made him understand the form and material of the brick as well as its potentials for emplacements, dependent on how it was carved (Marchand

2001: 160). For Ronni—in the same way as Merchand's relation to the brick—his relation to shellfish transformed into one of understanding rather than unfamiliarity. Ronni not only learned something about shellfish, he also embodied practices surrounding taste that made taste transformable through interaction. Experiences and encounters with ingredients and people shape how Ronni unravels tastes in the kitchen environment in which he is enmeshed as a chef. Tastes emerge through experiential phenomena and you come to understand them and master their production by both tasting and cooking.

When unravelling taste, such as the case of Ronni's change of attitude towards shellfish, one needs to overcome the former estrangement of the taste, which in many respects amounts to the non-existence of taste potentials. To underscore the estrangement–familiarity relation to taste it is useful to turn to what seems missing in this account of the embodied nature of knowledge through practice, namely failures. What do failures, breakdowns, and discontinuities in understandings of taste or even contested existences of taste, mean for a chef's relation to taste?

Collective Taste

Discrepancies between ideas of taste are very much to the fore when chefs engage in tastings together and especially when they have to work together in the kitchen, whether in the head chef–apprentice relation or in any other social relation. Discontinuities do not necessarily create arguments between chefs because they often think of the process of breaking with former ideas of taste as a process of learning. But at times the struggle to understand the strange becomes too much. To illuminate this, we will have to look at one simple dish whose taste requires a specific order of methods and a certain speed of execution. It is a small celeriac samosa that is filled with cold cream cheese and mackerel rillettes. On top, there are small dots of lemon gel, garnished with a piece of dill. This specific dish illuminates just how much your ability to understand and execute specific tastes, is central to your fitness within a kitchen:

> One day Simon, the youngest apprentice chef, had been put at the cold station in the kitchen. As I entered the kitchen I felt like something had dampened the otherwise cheerful atmosphere that normally permeated the air at around two in the afternoon. I followed Mikkel out into the yard, where he goes to smoke. He told me that Simon, the youngest apprentice chef, had broken down in tears the day before because Mikkel had told him that he worked too slowly with the celeriac samosa, which affected the ingredients and ultimately the taste—because of the different temperatures. The pickled samosa and the rillettes would blend together in a bad way and it would taste fishy and the dish would lose its freshness. Mikkel had tried to tell Simon how he should change his handling of the ingredients, but Simon did not agree or understand, so he had left the kitchen in distress, and this had caused the moody atmosphere. Mikkel reminded himself and repeated to me that each chef has "to learn to *work as we do*, or else the kitchen can't make tasteful dishes for the guests."

Simon returned to the restaurant kitchen the next day, and did not seem distressed about the situation, and the usual cheerful atmosphere of the kitchen seemed to have returned. I followed Simon into the cold section of the kitchen to hear his version of the story:

JØR "I hear that it was rough last Saturday . . ."

Simon "Well . . . Yeah . . . I work too slowly. I need to step up my game in this kitchen, I see."

JØR "But what happened?"

Simon "I think a lot of things went wrong . . . I was supposed to make samosas and I did not really know how to make them so I kind of just tried something out, which clearly wasn't right . . . I think that I just found out how *unaccustomed* I am to many of the things that happen in this kitchen . . . *I just felt like I didn't fit in*. Does that make sense?"

By recounting these events, it becomes clear how important it is to adapt to the rhythms of the kitchen when working as a chef. As Mikkel pointed out, you have to "work as we do" and at this point Simon represented a social position and an experiential position as someone who was not yet adapted to the kitchen environment and its tempo, cooking techniques and so on. Two weeks after I finished my fieldwork in the restaurant kitchen at Restaurant Mellemrum, Mikkel informed me that unfortunately Simon had had to quit because he did not feel he was capable of working as a chef.

To Mikkel, the fresh taste of the celeriac samosa was a matter of carrying out tasks with a certain speed of motion, and in the kitchen certain rhythms meant certain tastes. Therefore, Simon's slow engagement in the kitchen not only meant that he carried out tasks the wrong way, but it meant that he did not understand the work rhythms necessary to producing certain tastes. To Simon, the relation between tempo and taste outcomes was unknown. In Haydar Al-Mohammad's view, such a situation where the individual is in conflict with a social environment because of his or her lack of familiarity with certain issues, comes from a feeling of unhomeliness (2013: 220). In his article "Ravelling/unravelling: Being-in-the-world and Falling-out-of-the-world," Al-Mohammad argues that Martin Heidegger's notion of "being-in-the-world" does not mean that one is just part of the world per se; rather "being" means being entangled in the life-lines of other lives and by entanglement, one becomes part of specific everyday practices and not others (2013: 212). Thus, being-in-the-world means being part of specific social environments, falling out of some and never being part of others. According to Al-Mohammad, these entanglements do not come without work. One becomes entangled in the lives of others through everyday practical immersion, which is based on *practical wisdom*. Practical wisdom is a notion that Al-Mohammad has derived from Aristotle's concept of *phronesis*, which refers to a person's ability to comport with the surrounding world. Practical wisdom then, according to Al-Mohammad, determines the degree to which a person is capable of attaining a feeling of fitting into the social world, which is conceptualized as homeliness. Thus, homeliness is when a person's practical wisdom resonates with the everyday practices and know-how of the people and materials you are working to be "with" (Al-Mohammad 2013: 212). But, as Al-Mohammad remarks, even when homeliness might appear to have been achieved, it is continually at risk because you have to assert yourself if you are not to be forgotten in the social environment (Al-Mohammad 2013: 225).

As for Simon, even though he was established as an apprentice chef in the kitchen, which gave him a hinge in the social milieu on a formal level, he encountered a discrepancy between his practical wisdom and the know-how that was established in the kitchen. Al-Mohammad's notion of the way that one can go from being-at-home to not-being-at-home, corresponds with the way in which Simon was slowly disentangled from the kitchen because of his inability to fit in.

In the cases of both Ronni and Simon, their connectedness to certain social environments with an emphasis on taste, was at stake, but how can social connectedness be based on accordance with, or difference in, taste perception and practice? According to Antoine Hennion, one can describe the social arenas where certain tastes are accepted and others are not as *collectives* of actively tasting individuals (Hennion and Teil 2004; cf. Hennion 2006). To Hennion, the activities that unfold within the collective are what slowly but surely sets up a framework of comportments of material and social practices on which social communion relies. The chef would in many respects correspond to what Hennion calls the *amateur*, who knows how to comport himself within the taste collective, and the apprentice chef corresponds to the neophyte, who struggles to conform to the practices of the taste collective despite its possible incoherence with the apprentice's preconceived ideas of taste (Hennion 2007: 4). As the apprentice chef exercises taste practices among the other chefs, he is constantly reminded of the framework within which he needs to comport himself and his sense of taste. Thus, the expertise of the chef or the amateur does not rely on the amateur's sole expertise within a given field, but depends on his ability to resonate with the taste collective in which he resides. As Hennion and Teil argue, when social arenas are characterized by a focus on taste, the taste collective has the capacity to make one "depreciate what one loved, and to love what one despised," because the individuals belong to the collective (Hennion and Teil 2004: 31).

Conclusion

The kitchen at restaurant Mellemrum is a specific field that is affirmed through a framework of practices (production methods, handling of tools, aesthetic mechanisms, ethical values, memories etc.), which underscore the processes in which taste is being negotiated. And as Hennion points out, in a delimited field—such as a restaurant kitchen—the framework that exerts understandings of taste is not produced by a horizontal exercise of power. People have social positions within the kitchen, and this positioning allows some people to act as models, forcing the other members of the collective in specific directions with the force of the collective framework. These positions are evident from the fact that it is not the head chef, Mikkel, who has to leave the restaurant kitchen, but the apprentice chef, Simon. Thus, where Ronni's newly discovered relations between shellfish, cognac and sweetness made him enter more deeply into a social milieu, Simon, unfortunately disentangled himself from the taste collective that he had at one point wanted to become part of. The estrangement that Simon felt during the days of distress points to the fact that even though taste can be said to be something which is not inherent in the individual, individuals have different ways of negotiating taste in specific environments. As the scene with Simon shows, enmeshment is a struggle, because it means that you have to fight to feel "at home" in something, which at times might seem strange and unfamiliar. In the case of Ronni's understanding of the taste of shellfish, and in this case where Simon feels unfit for the kitchen, the issue of entanglement and disentanglement is at centre of these processes, and each chef's ability to resonate with their surroundings, practically, comes to define the degree with which each chef belongs to the taste collective that is established in the restaurant kitchen.

The professional restaurant kitchen provides a pivotal understanding of food taste as an indubitably social phenomenon that is transmitted between people and binds people to each other and the environment. Taste is enmeshed in the social milieu in which persons taste and make tasty things. Especially in the restaurant kitchen, the individual's personal preferences are not central when taste is being assessed; rather the social context in which taste is emplaced is the basis for structures that ultimately form taste practices. Chefs' work with taste shows how taste is a phenomenon that reflects the importance of the making of social environments and experiences.

Professional chefs act as windows into understanding the nuances and relativism of taste exactly because they cook food. Professional cooking is seen to be more than just an elaboration of ingredients; rather it is an activity where taste is culturally, materially and sensually shared and enacted. What these chefs teach us, is the fact that food taste is a phenomenon that links sociality and materiality in a forceful collaboration between experiences, taste collectives, and the qualities of ingredients. In relation to chefs, taste acts as one important aspect of the social world, which binds human beings together through materials.

References

Al-Mohammad, H. (2013). "Ravelling/Unravelling: Being-in-the-world and Falling-out-of-the- world," in T. Ingold and G. Palsson, eds., *Biosocial Becomings*, Cambridge: Cambridge University Press.

Argenti, N. (2002). "People of the Chisel: Apprenticeship, Youth, and Elites in Oku (Cameroon)." *American Ethnologist*, 29(3): 497–533.

Barnard, A. and J. Spencer. (2008). *Encyclopedia of Social and Cultural Anthropology*. London: Routledge.

Beriss, D. (2006). "What to Eat After the Storm, Restaurants and Hope, Part Two." *Anthropology News* 47(4): 49.

Berris, D and D. Sutton (2007). "Restaurants, Ideal Postmodern Institutions," in D. Berris and D. Sutton, *The Restaurants Book: Ethnographies of where we Eat*, 1–13. New York: Berg.

Bourdain, A. (2000). *Kitchen Confidential: Adventures in the Culinary Underbelly*. London: Bloomsbury.

Bousfield J. (1979). "The World seen as a Color Chart," in R.F. Ellen, D. Reason, eds., *Classifications in their Social Context*. 195–220. London: Academic.
Brenner, L. (1999). *American Appetite: The Coming of Age of a Cuisine*. New York: Avon Books.
Byrkjeflot, H., J. Strandgaard Pedersen and S. Svejenova (2013). "From Label to Practice: The Process of Creating New Nordic Cuisine." *Journal of Culinary Science & Technology*. 11(1): 36–55.
DeSoucey, M. (2010). "Gastronationalism: Food Traditions and Authenticity Politics in the European Union." *American Sociological Review*, 75(3): 432–455.
Ferguson, P. Parkhurst (2011). "The Senses of Taste." *The American Historical Review*, 116: 2: 371–384.
Fine, G.A. (1996). *Kitchens: The Culture of Restaurant Work*, updated ed. Berkeley, CA: University of CA Press.
Fine, G.A. (2009). "Introduction to the Introduction," in *Social Psychology Quarterly*, 72(3): 206.
Hennion, A. (2006). "Pragmatics of Taste," in *The Blackwell Companion to the Sociology of Culture*. Oxford, Blackwell: 131–144.
Hennion, A. (2007). "Those Things That Hold Us Together: Taste and Sociology." *Cultural Sociology*, 1:1: 97–114.
Hennion, A. and G. Teil (2004). "Discovering Quality or Performing taste? A Sociology of the Amateur." In: *Qualities of Food: Alternative Theoretical and Empirical Approaches*. M. Harvey, A. McMeekin, A. Warde eds, Manchester, Manchester UP, 2004: 19–37.
Hermansen, M.E. (2012). "Creating Terroir: An Anthropological Perspective on New Nordic Cuisine as an Expression of Nordic Identity." *Anthropology of Food* [Online], S7, Online since December 22, 2012, connection on October 12, 2017. www.http://aof.revues.org/7249.
Herzfeld, M. (2003). "The Body Impolitic: Artisans and Artifice in the Global Hierarchy of Value." Chicago: University of Chicago Press.
Hill, J. and D. Plath. (2006). "Moneyed Knowledge: How Women Become Commercial Shellfish Divers" in J. Singleton, ed., *Learning in Likely Places: Varieties of Apprenticeship in Japan*, 211–225. Cambridge: Cambridge University Press.
Howes, D. (2012). "The Cultural Life of the Senses." *Postmedieval*, 3(4): 450–454.
Leer, J. (2013). "Madlavning som maskulin eskapisme: Maskulin identitet i The Naked Chef og Spise med Price." *Norma*, 8(1): 43–57.
Marchand, T.H.J. (2001). "Minaret Building and Apprenticeship in Yemen." Richmond: Curzon.
Risbo, J., O.G. Mouritsen, M. Bom Frøst, J.D. Evans and Benedict Reade (2013). "Culinary Science in Denmark: Molecular Gastronomy and Beyond." *Journal of Culinary Science & Technology*, 11(2): 111.
Risvik, E., U. Larsson, R.E. Vatvedt Fjeld, H. Osa and M. Hersleth (2009). "Nordisk mat og språk. Å formidle det nordiske kjøkken—Et inspirationsnotat" ["Nordic food and language. To present the Nordic Kitchen—An inspirational note"], retrieved from http://nynordiskmad.org/fileadmin/webmasterfiles/PDF/Projekt/
Sutton, D. (2010). "Food and the Senses." *Annual Review of Anthropology* 39: 209–223.
Trubek, A.B. (2007). "Tasting Wisconsin: A Chef's Story," in D. Berriss and D. Sutton, eds., *The Restaurants Book*, 35–47. New York: Berg.
Warde, A. (2014). "After Taste: Culture, Consumption and Theories of Practice." *Journal of Consumer Culture*, 14(3): 279–303.
Visit Aarhus (2015). "Aarhus i den gastronomiske verdensklasse," retrieved from www.visitaarhus.dk/aarhus/aarhus-i-den-gastronomiske-verdensklasse.

11

Teaching to Cook and Learning to Sense in Food Education

Amy Trubek and Maria Carabello

Introduction: Bringing Our Senses to the Table

Taste is never simply or merely an internal subjective state, divorced from the broader context of our social lives. In fact, Lahne and Trubek (2014: 129) argue that, when tasting food, our "sensory experience *is* social experience." In this cast, the bite of cheese or the sip of wine involves physiological and contextual cues; there is always a dynamic between the intrinsic aspects and extrinsic values of the food or drink. This occurs both as a push and a pull; as Adam Chau (2008: 500) wrote, "human sociality is fundamentally sensorial." We taste in context and social contexts shape how we taste. When considering all the contexts involved, especially the making of the cheese as well as the assembling of the salad, and the multiple steps necessary to put together the casserole, another dimension of this constant interplay between sensory and social emerges. In order to be able to not just eat the cheese, salad or casserole but to savor the tastes, aromas and textures, we rely on our senses all along the way. When a dish is prepared, the cook is not required to engage with the sensorial realm, and yet it happens. If we want to have our food taste a certain way, we need to make it so. As Emily, a young woman interviewed about learning to cook said, "How do you learn what happens when you add mint to whatever unless you add mint to whatever? You're just never gonna know." So, how does it happen? And why?

Paying attention to sensory engagements is relatively new territory; food scholars have only just begun to tap into the myriad ways in which our social and sensory experiences are intertwined. The mutuality of the sensory and the social—the small gestures and grand assumptions involved in mediating our tastes, be they cultural or physiological—is ripe for more scholarly inquiry. Although food is often interpreted as a social good, a material object with social functions, the significance of the senses to its social nature (and also vice versa) can be neglected. This is often not a willful omission but rather a result of focus on the product—the finished dish—rather than the process—choices and actions to make it. For example, looking at a dish like ratatouille—in essence, a French vegetable stew that takes advantage of the aromatic herbs and vegetables that grow so well in Provence—we might recognize it as a product of the relationship between cooks and their natural environments, or view it as an expression of culinary identity due to an affinity to their geographic region. And often this is where the analysis ends. But what about more focused attention on the *steps taken* to make that dish? Might that help to bring the sensory to the fore? A focus on process as well as product—the discussion of how a bundle of herbs infuses flavor into the dish, or debates between cooks as to the importance of a particular variety of eggplant to achieve a soft texture, or the appropriate use of lemon to add acidity, or whether to chop or tear delicate herbs used as garnish—adds substance to the analysis. If such matters are not considered, this mutuality can get neglected and thus a certain depth and breadth missed. First, how do all of these minute sensorial decisions reveal the social identity and intentions of the cook? Second, how do her social conditions shape her sensory evaluations as she cooks and eats? Thus, in the *process* of creating this seemingly simple vegetable stew, broader sensory and social narratives about culinary identity and capacity are expressed, through communication, production and consumption.

The contributors in this volume all take seriously the relationship between the sensory and the social in order deepen and broaden the scholarship on food. In this chapter, our focus is on the social transmission and social imports of sensory knowledge; the means by which the taste, flavors and aromas of the eggplant, tomato, lemon and the bundle of herbs become accessible and conscious, and therefore part of a person's repertoire when they cook and eat. Such a repertoire emerges through dialogues with others, often in everyday contexts and conversations. Also, this repertoire plays a role in how these meals are shared and interpreted. We go on to argue that sensory experience and social interaction—for both scholars and cooks—are central to our relationship to food, and as food scholars, we might make certain moves; one, to acknowledge the centrality of the process, and the conditions of such a process, to making sense of food and drink, and then, next, to embed the dialogues and the actions intrinsic to this process into our own related pedagogy. Thus, the results of such a process are conceptual and tangible; for this is how sensory repertoires are produced and reproduced. Our explanations are shaped by ethnographic research on the cooking skill and knowledge of Americans (primarily in the Northeast region). Initially, this research took place with home cooks (Trubek 2017); in a subsequent project the insights gleaned were instrumental in the development of a cooking pedagogy that embraced the senses in teaching about the processes of making, eating and analyzing food.

Cooking and Sensing

For the past decade, we have been involved in a long-term, collaborative, transdisciplinary research program looking closely and carefully at cooking practices in the modern United States (see Trubek et al. 2017; and Wolfson et al. 2017). In the early phases of this work we visited twenty-five home cooks in rural, urban and semi-urban locales across the Northeast United States. We recorded one or two videos of each participant as they prepared typical dinnertime meals of their choosing from raw ingredients to finished dish(es). The age of participants ranged from the late twenties to early seventies. As part of a later study, five more home cooks were interviewed and videotaped (in the same region). At the conclusion of all the research phases, thirty home cooks had participated, generating thirty open-ended interviews and over fifty hours of ethnographic video footage of home meal preparation activities.

The video ethnographies revealed that all aspects of the process of making a meal—from choosing the right ingredients, to stirring and smelling, to tasting for enough seasoning, to biting to test for softness—involved embedded forms of sensory knowledge. For example, in one of the filmed segments, Karen—a middle-aged home cook living in a rural area of Vermont—taste-tests her boiled pasta to ensure proper tenderness. As Karen explains, this is "the Sicilian way" to test pasta. Having fished out a noodle and approved of its texture, she drains the pot and adds the pasta to the pan of simmering scampi sauce. While doing so, she explains that adding the pasta directly into the saucepan allows it to "get the whole aroma," which she clearly connects to the expected taste, "yummy." To be sure, though, she again tastes a forkful of noodles to make sure the dish is "fit for human consumption." She justifies tasting as a part of her process, explaining that when it comes to cooking, "I always play it by ear, by eyeball." She attributes the need to taste and smell to the ad-hoc nature of her practice, but perhaps the senses always emerge, *in situ*, as the best way to make a palatable dish.

In this clip, Karen—as with other home cooks—clearly demonstrated the role of the senses in guiding and informing the moment-to-moment decisions that arise throughout the cooking process. Other scholars have recognized this theme in their own fieldwork, leading to growing consensus that there is a critical sensory aspect to many skilled practices—for example, artisan cheese-making (Paxson 2012; West 2013), and more recently, cooking in general (Sutton 2014). For us and for them, focusing on sensory experiences indicates that a number of discrete component actions—for example, cooking skills, techniques, and strategies—share a sensory dimension; relying on vision, taste, texture and sound to facilitate the actions involved in preparing a dish. Thus, in our view, crucial to the process of learning to cook is the development of what Heather Paxson (2012) articulates as *synaesthetic reason*—that is, relying on multiple sensory domains to fulfill culinary aspirations. To us, this concept helps explain the fluid nature by which cooks like Karen process the stimuli of multiple sensory inputs to make decisions about their dishes, while also relying on one form of sensory input—for example, the smell of a sauce—to access other sensory expectations—for example, the taste and texture of a properly cooked pasta dressed with that sauce.

As we witnessed our informants' synaesthetic relationship to their cooking, we also realized that these complex sensory engagements often went largely undetected by the cooks themselves. That is, these sensory aspects were implicit in their actions and comments, but perhaps not at the fore of their articulated approach. Somehow the sensory dimensions of their process did not move past action to awareness; sensory omissions can easily occur with cooks and food scholars. Due to our video methodology, we were able to take note of hidden gestures, as we watched and re-watched the cooks at work. In the ethnographic videos we witnessed the cooks actively deploying their senses as they cooked; however, in our more formal interviews and even informal discussions, sensory deployments were rarely ever articulated as central to the cook's practice. People's focus tended to be on following recipes (or not), sourcing ingredients, and juggling multiple duties.

In addition to noting the unsubstantiated role that sensory knowledge had in many of the home cooks' processes, a connection between explicit sensing and cooking confidence was also made when analyzing the three most novice cooks who took part in our study. Reflections on learning how to cook, in any setting, always involved an opening, the sharing of the cook's realization that taste, smell and texture mattered—in the process as well as the outcome—as much as bodily nourishment. One of these cooks, Jen, reflected upon her earliest experiences learning to cook:

> [I learned to cook in] my junior year of college, which is when I moved off campus . . . yeah, it was pretty bad. I would say almost everything I ate that entire year was frozen, or from a box, like almost exclusively everything . . . At least 90 percent or more, or was to-go . . . And it was kind of interesting, because at the time I didn't think that was that bad, I [would think], "oh, that's what people do, they just buy a bag of frozen peas, and then throw some on a plate and put it in the microwave or whatever . . ."

While at the time Jen's social environment gave her little reason to second-guess her approach to meal preparation, as she grew older, she began to reflect more critically on her food practices:

> As an adult I felt more like, "geez, I never knew how much went into this whole damn process!" of eating, and cooking . . . I think that's where the 'I'm-definitely-not-an-experienced-chef-type' piece comes from, because I think I'm learning a lot, and I try to make decent meals, but I don't have a lot of background, it was just not part of my growing up. But I feel like I have done pretty well. I'm not opposed to being in the kitchen, but I definitely don't love it . . .

In a case like Jen's where she was never inculcated with the basic guidelines of meal preparation in her childhood home, she has looked for cues (that's what other people do) to help her figure out how she eats and prepares her meals, and also how she evaluates these practices. Yet her motivation to grow and diversify her culinary skill set came not just from social cues, but sensory ones. In our interview, Jen described her inspiration for learning to cook the burgers she prepared during the research visit:

> Nobody taught me how to make [burgers] actually, my parents don't make burgers like that. And what's kind of interesting is I don't even know where I came up with it . . . I think I just randomly one day started putting things into the burger mix, because my parents made really, really bland food. Growing up we would have, like, a piece of chicken on the grill, but it would just be like plain chicken on the grill, and maybe there'd be barbecue sauce on top, but that was it. And then it'd just be like plain mashed potatoes with butter and salt and pepper, and then like, plain green beans, with butter and salt and pepper. And that was like every night, or like steak and a starch and a vegetable.

Interviewer: So not a lot of seasonings beyond the butter, salt, and pepper?
Definitely not. So I think that, for me, I [thought], "I wanna just try and throw other things into this ..." And then I did, and they tasted decent, at least I thought so.

Learning to cook is not—and, never has been—a passive transmission from generation to generation, but always an actively acquired practice that involves exposure to some knowledge source, and the repetition of practice that allows the novice to incorporate various skills, techniques, and strategies into his or her culinary *modus operandi*. As anthropologist Tim Ingold (2013: 1) describes, "[t]o know things you have to grow into them, and let them grow in you, so that they become part of who you are." The mode by which this "knowing" happens in the culinary context, however, is far from universal. It can happen as children take an interest in the food work of their parents; it can happen as a dishwasher closely observes the symphony of the restaurant line at work; or, for cooks like Jen, it can happen more haphazardly. But whatever the context, every cook learns from someone else. As scholars and as educators, we were struck by the potential of making such sensory engagements *explicit*. Dialogue and actions that heightened the value of learning to sense could aid people as they learn to cook. Sensory articulations could help with solving problems, making in-the-moment decisions, and sharing kitchen successes and struggles with others.

Learning, Cooking and Sensing

Our ethnographic insights into the mutual importance of learning to sense and sensing to learn led us to embed both the social and sensory elements involved in cooking practice into our pedagogy. We are thereby able to enhance the sensory repertoires that are clearly crucial to understanding the multifaceted dimensions (both conceptual and tangible) of food. At the University of Vermont, we have taught food studies courses using a foods laboratory for over a decade (the serendipitous result of Trubek being hired to teach in the department that formerly housed a major in Home Economics). The foods lab has eight kitchen stations (with two students sharing each station), a large classroom/tasting room, a demonstration station and plenty of equipment. Over time, we developed a pedagogy for courses in the foods lab that intentionally aligns with certain principles of John Dewey's educational philosophy; that all aims of education should be instrumental in order to facilitate a sense of agency, both in learning and in life. Dewey and his colleagues at the early twentieth century Dewey School believed cooking, eating and conversing together created a type of democracy. We agree with this focus on community, and have combined the learning by doing approach with principles from vocational culinary education, which focuses on mastering skills (see Trubek, Belliveau, and Mares 2016). Our fidelity to learning by doing meant we diverged from traditional home economics curricula, especially in our emphasis on active engagements of all types, especially sensory ones. At this point, the curriculum has two central principles that differentiate it from others that are used in similar university settings: first, a commitment to encouraging the cultivation of "embodied knowledge" and making such knowledge the objective; and second, to instruct using inquiry based on a constant dialogue—between teacher and student, between student partners, between student and tasks—as the underlying educational philosophy.

Building upon the insights of our fieldwork, we have made the sensory *explicit* in our pedagogy. With every course using the foods lab, we try to create an environment in which students are encouraged to develop a sensory vocabulary, to describe and reflect upon their cooking process in sensory terms, and to have sensory discussions with their peers and instructors as they sit down to taste and share their dishes. By placing academic import on sensory knowledge, the students are led to realize that knowing food is not just about identifying ingredients, or memorizing principles of food science; knowing food is also *sensing* food. It is being able to communicate and act upon sensory cues and insights. As such, our approach is based on inquiry; we ask students questions like: What are the tools required for taking products of the natural world and making them palatable? How do you use those tools and to what ends? Why does consistency in the shape of ingredients have both aesthetic and sensory consequences? What is taste and flavor? How do you develop sensory attunement towards food and why? So, as the students learn to sense in the kitchen, they come to better know their food, they begin to grasp and anticipate the consequences of their cooking actions. Knowing and doing become intertwined.

Learning to Sense in Order to Become a Cook: A Study of Our Pedagogy

After the ethnographic research on home cooks was completed, our research program shifted; we wanted to take the insights we had into the culinary skill and knowledge of American home cooks and further apply them. Thus, beginning in 2014, we focused more closely on the foods laboratory pedagogy we used to teach our students. We had a coherent case for our multisensory and experiential teaching philosophy, but we wanted to test it, to enact it. To do so, we began to research the process and outcomes of our educational approach. We adopted similar methodologies to the previous project; we used interviews and videotapes to capture both perceptions and practices. The focus was on a course that we had taught multiple times before, always using the foods lab. There were thirty-two students enrolled in this annual Food and Culture course; broken up into two labs of sixteen students each. The lab experience integrated learning the cognitive, organizational, and technical skills involved in cooking with the application of concepts learned in lecture. Eight students, with different baseline levels of cooking skill, confidence and knowledge participated in an in-depth study analyzing the foods lab pedagogy. Each of these eight students was videotaped cooking in the lab on six separate occasions over the course of the semester. They were also interviewed at the beginning and end of the course to discuss and reflect upon their overall experience and growth.

Other researchers have noted that college students, on the cusp of independent adult life, have much to learn when it comes to cooking (Levy and Auld 2004: 200). This is especially true in the Northeastern United States, where a certain Yankee reticence has dominated the larger food culture. In this context, talking about food, especially as a bodily experience is traditionally dismissed, thus leaving many young adults without exposure to the intersections of making, eating *and* discussing dishes and meals. Young adult Americans, such as college students, must *learn* to smell, taste and share their experiences. The in-depth analysis of the video ethnographies allowed for unique "close readings" of the students' weekly trials, a witnessing of their efforts to pay attention to the cooking process, the appearance of the finished dish, as well as to the overall sensory qualities of the foods they prepared and consumed. The videos also captured certain dialogues, as the lab pairs talked to each other about what they knew before the course began and what they did outside the foods lab in their own home environments. We wanted to explore how students socialize through shared sensory experiences, but also how they are socialized into acknowledging, distinguishing, and paying attention to certain sensory experiences and what they communicate about the process of making a meal.

Learning from sensing and sensing to learn emerged as fundamental to the students' experience over the course of the semester. For example, after the course was completed, Claire identified "sensing" as a part of the process of making and eating meals that she would like to further improve:

> I think I'd like to improve more of my sensing … or understanding when something is done, like "how brown do I want it?" and how I want the final dish to look. More like how you want it to come out and taste, more so than just the process of it. Because I still feel like I'm focused on the process and not the final dish so much, and I would like to understand, "these are the flavors that I'm looking for in this dish." "I want it to taste smoky, or I want to taste more onion, or I want more sweetness, …" I just know that I want to have more of a focus on the final dish.

Claire differentiated between the act of cooking and the results of such labors, using sensory attention as a means of translating what she now knows about tastes, flavors and smells into a more desirable final product. While learning in the structured environment of the foods lab, Claire realized that to obtain confidence as a cook, she must consider the taste of the food, and also have *sensory aspirations*.

The Arc of Learning, Developing Synaesthetic Reason, and the Journey to Become a Cook

In the foods lab, students often do not begin with sensory aspirations; these develop over time, especially when nurtured in the learning context, through experience and conversation. Sensory knowledge has to be explicitly addressed by the instructor. In fact, the very first exercise done by students before they began working in the kitchen was learning the basics of sensory evaluation. Students are asked to do a guided tasting of several foods, some familiar, some not. They are introduced to the differences between taste and flavor (the former linked to taste buds on the tongue, the latter linked to interactions between taste, smell and texture), the five basic tastes (sweet, sour, bitter, salty, umami), the issues of temperature and texture, and more. It is only during the second week of the semester that the students tackled their first recipe: a basic potato, onion, and green pepper sauté.

Tasting food was an accessible first entry point, a way to ascertain, *in situ*, that students successfully integrated the knowledge imparted by the instructor. In one videotaped session of this lab, the first where students actively cooked by themselves—chopping, heating, cooking, stirring—Emily can be seen on video spearing a crisp potato with her fork, followed by soft caramelized onions and green peppers, as she assessed the outcome of her composed sauté. On the other side of the bench, Phoebe and Rachael appeared pleasantly surprised by both the taste and smell of their sauté. In the interviews, it came out that these sensory cues seemed to validate, even deem worthwhile, what Emily and others perceived to be a more laborious procedural approach followed in lab, which required cooking the various ingredients in a sequential manner, instead of the quick yet haphazard one-pan sauté method commonly used by the students at home (which they often described as "dump and stir"). As the camera shifted back to Emily, she appeared pleased by that first bite, prompting her to combine the two pans of ingredients to meld the flavors before plating and serving the final dish.

Visual aesthetics also factored into the students' process. Later in the lab, Phoebe began assembling a vegetable salad to complement the main dish. She meticulously arranged the carrot sticks and purple basil leaves around the edge of the bowl. With Phoebe and Emily, it became clear that while both were paying attention to sensory qualities in their dishes, Phoebe had yet to catch up to Emily in adopting a synaesthetic framing—that is, using one form of sensory input to make inferences towards the final experience of eating the dish.

During the making of momos (Nepalese dumplings) in the fifth week of the course, smell seemed to have become part of the open kitchen dialogue as Lucas can be heard favorably commenting upon the aroma of the ginger as he grates it. With another pair of students, it is not this pungent root that spurred dialogue, but the pan-fried bacon Brian opted to prepare as a communal garnish. Impressively, though, it was Eliza who remarked when she sensed the bacon aroma wafting over from the opposite side of the lab bench—perhaps, a signal that the class as a whole had become more attuned to sensory stimuli in their lab environment.

By the next week, the sensorium had become the foundation for Emily and Lucas's kitchen dialogue, as they now wanted to make sure the dishes were seasoned and harmonious before being presented at the table. As Emily stirred the ratatouille before tasting it, she removed the bouquet garni and jokingly remarked that this bundle of herbs and spices would not be suitable for eating. This prompted Lucas to chime in, "it would be good, if you could." This was clearly a comment indicative of a deeper sense of synaethesia, as Lucas had connected the favorable smell of the packet of herbs and spices with a desirable taste experience.

By the sixth week of lab, Eliza and Fern had also learned to communicate better with one another, and it was clear that the sensory inputs had also become central to their shared dialogue. Clearly approving of Eliza's vinaigrette, Fern tasted it and exclaimed: "pizzazz!" Later, Fern used her sense of sight to recognize the slight shimmer on the surface of the oil she added to a heated pan—"nice, it's a-ripplin'!"—a visible indication that the oil was hot enough for sautéing. Moments later, Eliza stated, "this is so good!," as she lifted the lid on the neighboring pan and was greeted with the sweet, savory aroma of onions, peppers, and the herbaceous bouquet garni. This, again, seemed a clear moment of synaesthesia as Eliza connected the smell of the dish with a prediction as to its overall quality and taste.

Across the bench, the scene of Brian and Claire spoke to the close relationship between sense and memory. Brian seemed at a loss for how to articulate the quality of his ratatouille; he was not sure whether it came out as it should have, because, as he put it, [I have] "nothing to compare it to." Claire, having made the dish many times before at home, had accumulated enough tacit knowledge through experience that when she tasted it she had the confidence to make adjustments. Eliza who, like Claire, came to the lab with confidence about the process of making this dish, made sure that both she and Fern had an opportunity to taste and season their rendition before calling it done. In these scenes, the discursive role of sensory memories emerged from the students' decision-making processes about how to adjust their dishes based on past experiences.

Two weeks later, the students switched gears from cooking to baking, as they prepared an apple galette with local Vermont apples. This was especially exciting for Emily who, in the interviews, talked at length about how she is far more confident and creative in her baking than in her cooking. Thus, as she knew, one key to baking a good pastry is starting with good ingredients, and she was audibly excited about the prospect of using Windfall Orchards apples—brought from the home of their guest instructor—in her galette. As the scene transitioned, she exclaimed, "oh my God, that's good!" noting a perfect marriage between honey and tartness in her bite of apple. This experience of tasting the apple in its raw state would also provide Emily a point of comparison once she later tasted the final cooked pastry. Across the bench, Phoebe seemed quite impressed by her pie dough, and synaesthetically connects its textural appearance with an expected mouthfeel: "flaky, yummy!"

By the final lab session, when instructors provided students with a mystery basket of ingredients and no direct guidance, the students were more reliant on their senses than ever before (and necessarily so). Emily tasted the sweet potato filling for her tarts before filling them, fully aware that adjustments to the taste and texture would no longer be possible once the filling had set in the oven. In the afternoon lab, Brian also made sure that he and his partner could easily taste throughout the cooking process by bringing over a set of spoons to their lab bench before he and his partner had even finished laying out all of their ingredients. By the time his final batch of corn mush fritters were fried and cooling, the level of seasoning and adjusting along the way seemed to have paid off:

Brian: <takes a bite of a fritter, and laughs to himself>
Lab TA: Are they good?
Brian: Yeah . . . they're [the] bomb!

The process of learning to cook has become a complex integration of sensory awareness and dialogue with a set of discrete mechanical (chop, stir, plate) and cognitive (plan, sequence, problem solve) skills. As this process emerged over the course of the semester, these cooks *enacted* what David Sutton calls "gustemology" (Sutton 2014), and they also *reinforced* how sensory experiences require social engagements. In the interviews the eight students who took part in the study recognized these transformations in their relationship to cooking, meal preparation and their culinary identities. Rachael began by reflecting on learning to cook at home when she was in high school:

> When I was getting older in high school my Mom wouldn't be home in the morning, so if I wanted breakfast I needed to know what to do. And then like, you know, if friends slept over and you make breakfast in the morning, just little things like that . . . I don't think I ever wanted to know how, it wasn't a, "I'm gonna sit down and learn how to make eggs," it was just out of necessity, and eggs were always around, and my Mom could make them, so it was on the mind if I wanted breakfast.

By the end of the semester, Rachael's perceptions of herself and her culinary practices had changed:

> *Maria (interviewer)*: You feel like you care more than you did at the beginning?
>
> *Rachael*: Before you asked about like an arc of change, that's another arc of change that like at the beginning I only cared because it was part of school, and then I cared because [of the instructor] teaching it . . . and then I only cared because it was really fun, and I wanted to eat, and then I really started caring because of like the bigger things that we were talking about in lecture. So, there was definitely a change of like, "I'm not just here cooking, I'm part of this, like, grander change now that I'm here." And so, and you start caring because you're cooking for your friends now, and then cooking for your Mom and there was definitely an arc of why I was there caring. I think I was always caring, but I don't necessarily care about cooking because I know that I'll eat, I won't go hungry, I know that. But, that you can care because it's part of this bigger picture.

Fern's appreciation of sensory complexity, the various tastes, flavors and smells that can be part of sensory experiences, was what she identified as transformative:

> *Fern*: Yeah, I normally just taste it for salt and pepper.
>
> *Maria*: Like, seasoning?
>
> *Fern*: Yeah, seasoning. That's really what I taste it for like the spices like curry or something or texture, like pasta for example. You need to know how hard it is [so] I taste it to try to figure it out. That's probably something I could have done more in lab, tasted and decided.
>
> *Maria*: Has the way you season your food as you cook at home changed at all?
>
> *Fern*: Yeah, I'm not as scared of curry and coconut milk, and all those different things. I'm more willing to try it. They're spices and they can be really good,

> especially yesterday, the curried cabbage was *so* good compared to the first time we did the curried cabbage, when I couldn't eat it. And then yesterday I just wanted more, so you know it's just that certain things make things really good if you just add to it and change it up.

For Brian it was his relationship to the food cooked by *others* (here, at a restaurant) that he noted to have transformed. He explained:

> [I am] definitely just way more aware of food, and um, especially like tasting it, and like if I go out to a good meal, you know?
>
> I try to appreciate it, and actually take the time to think about what the food tastes like, and I think more about ingredients, and I don't know . . . I was already thinking about food all the time, but now I feel like it's even more so, or on a deeper level.

Lucas's experience revealed that the mode of learning, the emphasis on sensory analysis through experience and repetition, was not necessarily teaching the students a precise recipe for achieving certain flavors. Rather, it was giving them the sensory toolkit and guideposts to develop as a cook, both inside and outside of the lab:

> *Maria*: Do you feel like you've gotten better as a taster?
>
> *Lucas*: Definitely with the herbs. With herbs and spices, I've definitely gotten a lot better, I mean not all of it, it was mostly because the lab sparked my interest and then I kind of looked more into it, and started paying more attention to it. So, it's not that I necessarily learned it from the lab, but it made me pay attention to it, which hopefully [will continue].

The social complexities involved in making meals is also identified as fundamental to the experience, for even a college student does not always cook alone or solely for themselves. The tensions between learning to be a competent cook and cooking with and for others represented a constant in their experiences in and outside of lab:

> *Emily*: Like you can sauté at home, it's kind of haphazard, you throw it together, who cares really if the pieces are uniform, because you're just gonna eat it, and that's dinner for Wednesday . . . But here [in the lab] when you're like, well what makes this dish harmonious? All the pieces being a similar size, so when you put a bite in your mouth, it's not like, "oh, I have a giant piece of potato and a tiny piece of pepper and a giant piece of onion . . ." It's, it's . . . everything looks similar and thus there's something more aesthetically pleasing about that. So I usually don't put that much care into my cooking [at home].

A lab setting offers fewer distractions, so sensory and aesthetic aims are easier to prioritize, but now Emily seeks to realize these aims both in the lab and at home. The lab setting created self-awareness, as well as new aspirations.

Conclusion

This chapter on experiential culinary education in the Northeastern United States traces the journeys of young cooks as they learn about cooking skills, techniques, and repertoires. These experiences—captured through interviews, video recordings, and kitchen observations—highlight the instrumental role played by *sensory attention* both throughout the young adults' cooking processes, as well as in their consumption of the products they prepare for and with others. The reciprocal quality of the sensory and the social is presented and then confirmed, since sensory attention is "a learned and active practice in which sensations arise neither from the food nor from the consumer, but from the encounter between them, that is, it is neither taste nor taster, but *tasting*" (Lahne and Trubek 2014; 130 cf. Hennion 2005, 2007; Teil and Hennion 2004).

The young cooks in this study treated sensory experiences with food as *active*, *acquired*, and *emerging from social contexts* again and again in the reflections we heard, and practices we witnessed. As Claire learned to envision and articulate her sensory aspirations for her cooking, Fern began to reach beyond salt and pepper when seasoning her dishes, and Brian started to meditate over the flavors on his plate while out at a restaurant, it becomes clear that these young cooks successfully immersed themselves in the reciprocal practice of learning to sense, and sensing to learn during their time in the foods lab.

Much like the struggle to legitimize a focus upon the senses as significant to food scholarship (Sutton 2010, 2014), in order for the students to transform their practices and develop their culinary identities they had to first open themselves to the idea that there is more than one way to learn, to know, or to practice food:

> *Emily*: I love the class, and I love talking about food issues and all this stuff, but I think there's also something so rewarding about going into lab, turning off the academic brain, and doing something that's sensorial and … not thinking. You're thinking, "how do I make this kale good?" It's a totally different experience. Does it need more salt? Does it smell good? You know, what would happen if I added this herb? That's a totally different way of working than thinking … theoretically about transcultural food practices. I think that's … it's a really good thing to do both, and I've enjoyed the opportunity to do both … It's not just cooking, it's like we're learning as we're doing it. Getting to sit down with people, and meet people in a so much more intimate way than most of the interactions you have at these universities … [was] pretty unique.

In this response, Emily explains how learning to sense and sensing to learn has enhanced both her curiosity around food and her everyday cooking practices, while also hinting at the promise of her future development as a cook and eater. More than promoting a mastery of any certain dish or recipe, it is this foundation of culinary curiosity facilitated through sensory attention and observation that we believe will encourage these students to continue to become more engaged, fluent, and persistent cooks long after their semester in the foods lab ends. The lab may have provided them with a scaffolding to guide their initial education, but following their senses and remaining curious every time they enter the kitchen to prepare a meal, these students will reaffirm the power of the senses and the centrality of social contexts to knowing about food, in theory and in practice.

References

Chau, A. (2008). "The Sensorial Production of the Social." *Ethnos: Journal of Anthropology*, 73 (4).

Hennion, A. (2005). "Pragmatics of Taste," in M. Jacobs and N. Weiss Hanrahan, eds., *The Blackwell Companion to the Sociology of Culture*. Malden, MA: Blackwell.

Hennion, A. (2007). "Those Things That Hold Us Together: Taste and Sociology." *Cultural Sociology*, 1(1): 97–114.

Ingold, T. (2013). *Making: Anthropology, Archaeology, Art and Architecture*. London: New York: Routledge.

Lahne, J., and A.B. Trubek. (2014). "'A Little Information Excites Us': Consumer Sensory Experience of Vermont Artisan Cheese as Active Practice." *Appetite*, 78 (July): 129–138.

Levy, J., and G. Auld (2004). "Cooking Classes Outperform Cooking Demonstrations for College Sophomores." *Journal of Nutrition Education and Behavior*, 36(4): 197–203.

Paxson, H. (2012). *The Life of Cheese: Crafting Food and Value in America*. Berkeley: University of California Press.

Sutton, D.E. (2010). "Food and the Senses." *Annual Review of Anthropology*, 39(1): 209–223.

Sutton, D.E. (2014). *Secrets from the Greek Kitchen: Cooking, Skill, and Everyday Life on an Aegean Island*. Oakland: University of California Press.

Teil, G., and A. Hennion. (2004). "Discovering Quality or Performing Taste? A Sociology of the Amateur," in M. Harvey, A. McMeekin, and A. Warde, eds., *Qualities of Food*. Manchester: New York: Manchester University Press.

Trubek, A.B. (2017). *Making Modern Meals: How Americans Cook Today*. Berkeley: University of California Press.

Trubek, A.B., C. Belliveau, and T. Mares. (2016). "Emergence and Repetition: Teaching Food and Culture Using a Foods Lab." *Journal of Pedagogic Development* 6(1): 23–30.

Trubek, A.B., M. Carabello, C. Morgan, and J. Lahne. (2017). "Empowered to Cook: The Crucial Role of 'Food Agency' in Making Meals." *Appetite*, 116 (September): 297–305.

West, H.G. (2013). "Thinking Like a Cheese: Towards an Ecological Understanding of the Reproduction of Knowledge in Contemporary Artisan Cheesemaking," in R. Ellen, S.J. Lycett, and S.E. Johns, eds., *Understanding Cultural Transmission in Anthropology: A Critical Synthesis*, 320–345. Oxford: Berghahn Books.

Wolfson, J.A., S. Bostic, J. Lahne, C. Morgan, S.C. Henley, J. Harvey, and A. Trubek. (2017). "A Comprehensive Approach to Understanding Cooking Behavior: Implications for Research and Practice." *British Food Journal* 119(5): 1147–1158.

Part Four

Taste Politics

12

Taste Activism in Urban Sardinia, Italy[1]

Carole Counihan

Introduction

This chapter explores taste activism—the purposive public mobilization of the senses to challenge the ills of the agro-industrial food system and promote local, sustainable, high-quality food for all. Taste activism uses sensuous experience as a catalyst and method of political action. It takes place in social spaces as people exchange food, knowledge, and sensory experiences, and use mind and body to develop a critical approach to food. I examine the deployment of taste activism in and around Cagliari, the capital of the Italian island region of Sardinia, a major Mediterranean port, and a commercial hub of 160,000 people.

After centuries of a largely self-sufficient agro-pastoral-fishing economy, since the 1980s Sardinia's food system has become increasingly industrialized and globalized. Supermarkets, processed foods, and long food chains have burgeoned, but so have oppositional alternatives like vegetarian/local restaurants, urban gardens, farmers' markets, farm-to-school programs, solidarity buying groups, and Slow Food—initiatives that I studied ethnographically over several months between 2011 and 2015. Many featured organized tastings as well as constant informal discussions of taste, which were important ways of educating about quality food and building resistant communities.

This chapter focuses specifically on one event—the first anniversary celebration of Slow Food Cagliari's founding of the "Selargius caper food community" to show taste activism in action. Capers are the buds from a Mediterranean shrub, which are preserved in salt, vinegar or oil and used in cooking. The Slow Food event had three components that worked together to engage participants intellectually, corporeally, and socially: a panel discussion about capers and their history, a walk to nearby fields to see and learn about caper bushes from a plant scientist, and a guided caper tasting. The chapter explores how these events relate to Geneviève Teil and Antoine Hennion's (2004) claim that taste is a holistic, interactive, social practice that can develop perception, reflexivity, and community. But it goes further in showing how taste can also be a tool for political action.

I participated in the Slow Food caper celebration as part of my ethnographic research on food activism in Cagliari and vicinity for several months in three visits between 2011 and 2015 (Counihan 2014a, 2014b). I conducted informal and formal recorded interviews with over seventy people and did participant-observation in farmers' markets, urban gardens, panel discussions, restaurants, meals, tastings, farm visits, and a wild-herb-gathering expedition. I had done previous research on Italian foodways in Cagliari in 1976, in Bosa (Sardinia) in 1978–1979 (Counihan 1981, 1984, 1990), and in Florence in 1982–1984 (Counihan 2004). Invitations from Professors Gabriella Da Re and Benedetto Meloni to be a visiting professor at the University of Cagliari in 2011, 2015 and 2016 triggered this ethnographic research project on food activism.

Taste activism is a subset of food activism, which, to use Slow Food's ingenious slogan, is the pursuit of "good, clean, and fair food"—tasty, culturally appropriate, sustainable, unpolluting food produced at fair prices for workers and consumers. I studied food activism for several reasons. First, it is growing all over the world in response to food insecurity, inequality, and unsustainability (Counihan and Siniscalchi 2014), and many efforts have emerged in Italy.[2] Second, cities are crucial to global food systems (Bosso 2016), and Cagliari is a revealing case study because its metropolitan area must feed almost 600,000 people, a third of the entire population of Sardinia. Third, compared to other EU countries and the United States, Italy experienced industrial production and large-scale chain distribution of food relatively late (Scarpellini 2004), and thus Italians still have living memories of local food (Counihan 2004). Last, like many Italians (Black 2012; Counihan 2004; Marovelli 2014; Morini et al. 2015; Ochs, Pontecorvo and Fasulo 1996), Cagliaritani are passionate about taste and strongly attached to their local cuisines, making a study of taste activism highly relevant to their own pressing concerns.

Taste Activism

This chapter contributes to the literature on taste by showing how in Cagliari, taste figures prominently in many efforts to raise consciousness about the food system and change behavior towards local and sustainable consumption. I define taste as the cultural, social, multi-sensorial experience of food involving flavor, smell, sight, touch, and sound.[3] Taste activism relies on what Sarah Pink (2009: 24) calls "embodiment"—using the body as "a source of knowledge and subsequently of agency." Teil and Hennion (2004: 19) develop the concept of taste as agency when they write: "Taste is an activity and not a passive or determined state." Taste is experienced and defined through relationships as people eat together, share food, and share their opinions about it; taste is a social connector (Teil and Hennion 2004: 25). Moreover, taste develops "the emergence of reflexivity," and "the practice of perception" (Teil and Hennion 2004: 20), which materialized in the Slow Food caper event discussed below.

The role of taste in learning, connection, and change is enhanced by its ability to carry emotions and memories, as David Sutton (2001) and Nadia Seremetakis (1993) have shown for Greeks. Italian food activists became emotionally engaged as they remembered, learned about, ate, and informed others about delicious local foods associated with home, place, culture, and identity (Counihan 2014a, 2014b). These emotional connections, enacted through taste and commensality, were essential to sustaining social movements, as Fernando Bosco (2006) argued about the Mothers of the Plaza de Mayo in Argentina.

At the caper tasting in Cagliari, participants displayed emotions in a process similar to what Jerome Bruner (1986: 115, quoted in Taggart 2012: 418) called "attunement"—the "mutually reinforcing, mutually confirming" process of caretaker and child responding to each other that is essential to the child's development. Anthropologist James Taggart (2012) applied Bruner's concept of attunement to the process of narration and response characteristic of Mexican Nahuat oral story-telling, which was an important way to construct emotional understanding. The discussion at the caper tasting—although more cursory and less intimate—involved a similar attunement as people's give-and-take established beliefs and feelings about the capers and local food more generally. This process helped tie people together into the network essential to building food activism.

I argue that taste facilitated political change because it inspired passion. It enabled what Sardinian-born Marxist politician and philosopher Antonio Gramsci called: "movement from knowing to understanding to feeling and vice versa from feeling to understanding to knowing . . . you cannot make history and politics without passion" (Gramsci 1975: 451–452, my translation).[4] Food via the senses and ingestion wove together mind, sentiment, and body; it enthused many and led some to oppositional politics. In the conclusions I also raise further questions about the long-term impact of taste education on changing behavior and the extent of critical thinking it launches.

Slow Food, the Selargius Caper, and Taste

The political potential of taste, its emotional resonance, and its holistic social construction were revealed in the first anniversary celebration of Slow Food's establishment of a "food community" based on the Selargius caper (*cappero selargino*) held at the Casa del Canonico Putzu in Selargius on June 12, 2011. The event centered on a group experience of learning, sensing, tasting, and talking about the special local product of capers, and demonstrated how people, places, production methods and processing techniques contributed to Selargius capers' unique taste; it then featured a guided tasting and discussion of the criteria that determined good and bad capers. The event gave participants a chance to develop their values and activism through experiencing and debating taste, and it revealed how Slow Food worked on the ground.

Slow Food is an international organization that was founded in Italy in 1989. It has approximately 100,000 members from 150 countries divided into 1,500 local chapters called convivia (*condotte* in Italy) led by chapter leaders who carry out various activities on a volunteer basis. Slow Food aims to promote "good, clean and fair food" by safeguarding endangered high-quality local products, supporting small producers, educating consumers, and building local and global alliances.[5]

One tactic it uses is to establish local "food communities" around specific heritage products like the Selargius caper. According to the Slow Food website, a food community is "a group of small-scale producers and others, united by the production of a particular food and closely linked to a geographic area. Food community members are involved in small-scale and sustainable production of quality products."[6] Their attachment to the product, its *terroir*, and its producers is strong. Good taste, holistically defined and continually debated, is central to Slow Food's determination of quality.[7] Taste education is an ongoing strategy to engage members, develop their powers of sensory discernment, and change behavior.[8]

Establishing the caper food community was a key accomplishment of Slow Food Cagliari. It was an active chapter with between 150–200 members and an organizing committee of nine core participants led by two-term chapter leader (*fiduciaria*) Anna Cossu whom I interviewed in 2011 and 2015. The chapter worked to safeguard special local products and organized visits to producers, educational events, tastings, and themed dinners. In 2010 Slow Food Cagliari established the Selargius caper food community to stimulate caper cultivation, help the local economy, and, as anthropologist and Slow Food Cagliari organizing committee member Alessandra Guigoni put it, "to preserve the natural, agricultural, and cultural heritage tied to this plant" (2010: 12, my translation).

Caper production is a microcosm of Sardinian agriculture. Today Selargius is a semi-urban community of 29,000 just eight kilometers from Cagliari, but it has a long history as an important agricultural area where grape and wine production dominated the economy until they were destroyed by the phylloxera pest that swept Europe in the early twentieth century. During that crisis, the caper economy kept people alive (Guigoni 2010: 10). Selargius women sold them in large baskets in the neighborhood markets of Cagliari up through the 1960s. Caper production fell along with the entire agricultural sector in the 1970s due to job opportunities elsewhere and competition from lower priced globalized capers. But after years of abandonment, in 2000 small farmer Marco Maxia was able to get them back into full production just by weeding and pruning. He was instrumental in inspiring others to tend capers and Slow Food to take an interest in their production. He was part of the panel discussion and caper tasting at the 2011 anniversary celebration, and he spent a day in 2013 showing me his fields and processing workshop and explaining all about capers.

The First Anniversary Celebration of the Slow Food Caper Food Community

The first anniversary of the founding of the caper food community was a tripartite learning experience that engaged intellect, senses, and sociability. Around fifty people participated in the afternoon's events, which began with a panel discussion on the history and current status of caper production in Selargius, was followed by a guided walking tour of nearby caper fields, and culminated in a caper tasting. The friendly atmosphere, collective learning, and sensory encounter with the growing plants prepped the participants for appreciating the taste of the capers.

Slow Food leader Anna Cossu opened the panel by welcoming everyone and briefly describing Slow Food's principles of good, clean and fair food and its support for revitalizing caper production. One of the things Cossu said was, "knowledge is the foundation of everything," a belief that clearly inspired the rest of the day's events. A range of participants spoke on the panel and revealed multiple perspectives about capers. Politicians from the town and province spoke about agriculture's deep roots and potential economic value to the area, highlighting the benefits of eating local. Caper farmer Marco Maxia said he rented his first caper plot in 2000 and endured eight difficult and discouraging years before he finally began to make a small profit. The biggest boost came from Slow Food, which got involved with capers in 2004 and invited Maxia to its renowned biannual food exposition Salon of Taste (*Salone del Gusto*). Caper farmer Efisio Secchi spoke with astonishment of how he had been picking capers since boyhood and never imagined that they would be valued enough to be featured at Slow Food events in Selargius, not to mention at the Salon of Taste in Torino.

Alessandra Guigoni recounted that the bush-type caper plant grows throughout the circum-Mediterranean region, in arid areas of north Africa, and the islands of Pantelleria, Santina, and Sicily, as well as Sardinia, where the first records of the caper's culinary and medicinal uses dated to 1725 (Guigoni 2010). Another panelist, local resident Stefanina Dentoni, spoke about her great-grandfather's role in launching modern caper production in Selargius in the mid-nineteenth century when he migrated there from Liguria near Genoa on mainland Italy. He eventually planted thirteen hundred caper bushes rendering over two tons of capers per year. Stefanina Dentoni talked about how hard the pickers worked and how they had to make sure there were no flowers on the plants at the end of the day because every flower was a lost caper.

After the panel, we all walked a few hundred meters down the street to a field where an expert botanist spoke about capers. People strolled around and chatted informally with other participants while examining the bushy bright green plants, touching the leaves, spotting the buds, and picking a few of the gorgeous flowers. We learned that capers fit the local soil and climate perfectly for they were sturdy and resistant to the wind and aridity typical of Sardinia, with roots that could reach ten meters deep or more. The plants could grow to a meter and a half in height and three meters in diameter and live over a hundred years. In Selargius, capers thrived in the context of extreme land fragmentation typical of Sardinia and were planted along roads, as land boundaries, with fruit trees, and near the grapevines that produced the vinegar used to pickle capers—which were also salted as the tasting revealed.

A caper (*cappero, bocciolo*) is the bud of the plant before it flowers. The smaller the buds, the more prized they are; the more they are picked, the more the plants produce. The harvest starts around mid-May and goes through mid-August. At the start of the season, the pickers go to each plant every five to six days, but as the days get longer and warmer they pick every two to three to four days. A bush can render from 700–800 grams per cutting per plant up to two to three kilograms per cutting of the biggest plants during the summer's longest days when they are most productive. The older and bigger the plant, the more it produces and the better the quality, but even new plants can produce in the second year. In contrast to past practice, today at the end of the season some plants are allowed to flower and produce the caper fruit known as *capperone*, which is pickled and commercialized, especially for cocktails, in an effort to diversify the plant's marketability.

The visit to the fields engaged all the senses except taste, which came later. The sunny and breezy late afternoon was beautiful and the walk to and around the fields was sociable. I chatted with former Slow Food Cagliari leader Anna Sulis, whom I had interviewed earlier, and mentioned that I was hoping to meet a butcher named Walter Vivarelli who sold only Sardinian heritage beef. Sulis immediately pointed him out and introduced us. We had an amiable chat and then set up an interview for the next day. Others were similarly making new friends, renewing acquaintances, and extending their networks.

The Caper Tasting

After an hour in the fields, we walked back to the Casa del Canonico Putzu for the caper tasting. It was led by a member of the Slow Food Cagliari organizing committee, forty-seven-year old Luca Galassi. In an earlier interview he had told me he was the chapter member in charge of conviviality, which was integral to every Slow Food event. After a day of learning, he said, it was critical "to bring the members together at the table to share this learning and transform it into sensory knowledge of the products." Galassi was adamant that the pleasures of eating and socializing were central to Slow Food's work and that: "we have to link whatever activity we do in Slow Food with the chance to know sensorially what we are talking about . . . Let's talk about a food, let's do a day on the caper, then let's eat the capers . . . The cognitive, cultural, solidarity aspects have to be accompanied by the tasting aspect." His approach reflected sensory anthropologists Howes and Classen's (2013: 3) claim that "ideas are communicated through sensory impressions all the time."

Galassi put into practice Slow Food's belief in corporeal learning by leading a caper tasting to inform people about several varieties of capers and their organoleptic properties, with special attention to the local ones. Outdoors at the Casa del Canonico Putzu, Slow Food volunteers had set several tables with capers arranged on paper plates, like numbers on a clock, for participants to taste. Each table of eight to ten people also had crusty *pistoccu* bread, bottles of mineral water, and baskets of cherry tomatoes donated by the tomato cooperative of S. Margherita di Pula (to which Slow Food Cagliari had organized a visit a few weeks earlier). Slow Food volunteers went around pouring wine, which people welcomed enthusiastically, and Luca Galassi led us through the tasting, which I digitally recorded, picking up Galassi easily and my table-mates with varying success depending on how close to the recorder they were.

I sat at a table with educator and former Slow Food chapter leader Anna Sulis; University of Cagliari sociologist Benedetto Meloni and two of his friends, Carmen and Giorgio; organic farmer/distributor Matteo Floris, his partner Paola, his mother, and one other woman. It was very festive and sociable, especially as the wine flowed. People chatted light-heartedly about the capers as they tasted them and tried to identify flavor differences and preferences.

Luca Galassi began the tasting thus:

> This is very simple. You have in front of you a plate with a cycle like a snail, which is the symbol of Slow Food. Your reference point is the hour twelve on the clock

> and from there we will proceed clockwise, tasting several different capers. We will make comments and present information, and after tasting the capers you can clean your palate with the water . . . On the plate you will find capers that are very different from each other in provenance and preparation . . . We begin with the Selargius caper, which you can see is much smaller than all the others, and if you try to squash it, it releases much less water than the others . . . When you taste it, first try to smell its fragrance and then put it in your mouth and taste it.

Galassi then guided us around the plate, having us taste next a Slow Food endorsed "presidium" caper from the island of Salina,[9] one from the island of Pantelleria, "a supermarket caper," one pickled in vinegar, one preserved in olive oil, and finally the caper fruit or *capperone*. The organization of the tasting allowed sensory comparison and evaluation of different products. It put into practice the belief that knowledge delivered socially and sensorially could affect people's experience of taste which, the hope was, would affect their subsequent purchasing and consumption.

Galassi encouraged a holistic experience of taste by incorporating not only people's olfactory sense and mouth feel but also their attention to visual appearance and texture. He said:

> Selargius capers are more puffed up and they have a more complete rotundity . . . Let's pay attention to the shape . . . They always have the stem, which is gone in the others . . . I invite you to feel it for a second between your teeth, how it is more crunchy (*croccante*) than the other capers that have absorbed so much water.

To bring in the producer's presence and perspective, Galassi asked Marco Maxia to comment. Maxia said:

> The Selargius caper has a much lighter specific gravity compared to other capers and in consequence it retains much less water. This is one of the characteristics contributing to its demise because having less water, it weighs less . . . and so for the same volume you have less weight and thus a lower price . . . But the particular contribution of this lightness is above all to flavor, there is a good flavor that you can taste.

The event gave caper farmer Maxia a chance to explain the particularities of his product and their impact on taste, enriching not only people's understanding but also their personal connection to him, potentially heightening their commitment to his product.

The tasting fostered connections not only between consumers and the producer but also among consumers. The event provided many pleasant opportunities for socializing. At my table the conversation was animated as people chatted about the different capers, which ones they liked and did not, how hard or easy it was to differentiate them, and especially about how salty they were. People actively participated in responding to the capers, to each other, and to Galassi's suggestions. They spoke with a kind of whimsical anarchy and playfulness as this excerpt from my recording reveals:

Galassi: Let's try another product, a very famous one, the Salina caper, which among other things is a Slow Food presidium.
Carmen: We visited Salina two years ago. There too Malvasia wine and capers.
Galassi: We have a particular fondness for this product . . . I ask you to proceed as before, that is, smell its fragrance and then taste it, after which I will make a few comments.
Carmen: They are very salty.
Giorgio: Tonight we'll have to sleep with a bottle of water.
Galassi: What is the principal characteristic of this compared to the other, what have you tasted more than in the earlier one?
People: Salt . . .
Giorgio: It has a taste, a strange aftertaste.
Galassi: Perhaps in terms of the flavor, if we manage to separate the salt from all the rest, we taste a perhaps more intense flavor, a stronger flavor?
People: No . . .
Galassi: Now let's turn to the caper from Pantelleria. Probably those of you not already fond of the Selargius caper have heard tell of the Pantelleria caper that is often presented with products of high quality and great interest. You can turn to that one . . .
Carmen: It doesn't taste of anything.
People: Chatting
Carmen: It tastes of salt, only of salt.
Man: Even more salty.
Carmen: Even more salty, bravo . . .

Clearly the strongest taste was saltiness which overwhelmed more subtle flavors and people voiced their opinions freely about it. The tasting was very informal, not an exercise in distinction as some tasting events can be (Meneley 2014). People's spirited conversations, only partially captured in the recording, were ways of communicating and debating not only opinions but also emotions through a process of attunement (Bruner 1986). James Taggart (2012) has pointed out that there is a difference between feeling something and representing that experience verbally, but words are the richest pathway we have to share and access each other's emotional experiences. Taste shares with emotion the difficulty of being put into words, but the caper tasting gave people a chance to practice externalizing their sensory impressions and comparing them with others' and they did not hesitate to express their opinions and disagree with the expert.

Galassi did not try to impose a taste hegemony although he did lead us to be critical of the mass-produced caper about which he said, "Let's turn to the ultimate dish of the evening. You will try an unspecified caper that we call a supermarket caper, which is the one we get when we have to make a recipe in a hurry. Try that one and then we'll draw some conclusions." After a few moments of participants' tasting and commenting, Galassi said, "I heard someone use the adjective 'terrifying.' These supermarket capers are absolutely different . . . They have a less pleasing appearance and less delicacy, let's get back to their form and pay attention to that." But at my table someone commented, "I can't tell the difference," showing the processual and interactive nature of attunement.

Galassi gently prodded people to be critical tasters by encouraging sensory awareness and group discussion, which built relationships with table-mates. For example, he said, "Let's try the caper preserved in olive oil. Does anyone have anything to say about this caper?" He listened to what people said and then repeated their comments to the whole group—fostering attunement. The tasting covered three kinds of preservation techniques—in salt, oil, and vinegar—and encouraged participants to evaluate their different flavors and mouth feel. Happily the last capers were more pleasing to most of our palates and comments included: "this is good, finally one that doesn't taste like salt, you can taste the vinegar," "it's good, really good," "it tastes of red pepper and garlic, the ones in vinegar are delicious." Galassi disseminated Slow Food's ideology and denigrated the supermarket caper, but participants expressed their agency by sharing a range of viewpoints. The give-and-take facilitated the "mutually reinforcing, mutually confirming" (Bruner 1986: 115) process of sharing and recognition typical of attunement.

Conclusion: Tasting and Agency

The caper tasting lasted about an hour and was a pleasant culmination of a day of sensory, social, and cognitive learning. The experience supported Teil and Hennion's (2004: 25) claim that, "taste is a way of building relationships, with things and with people." Moreover, the public interactive context encouraged the development of perception and reflexivity characteristic of taste underscored by Teil and Hennion (2004). Participants left the event knowing why Selargius capers were special, how they were uniquely suited to the local environment, how they were processed in salt or oil or vinegar, who the farmers were, and what challenges they faced in making a living from caper production. Successfully or not, participants tried to use their senses to perceive the unique taste, lightness, and delicacy of Selargius capers. They left the event with greater awareness, which could influence their subsequent food choices and political action.

In an interview, Cagliari chapter leader Anna Cossu described how taste education built critical thinking, as Teil and Hennion (2004) claimed:

> It has happened to me in our tastings that the tasting leader says, 'oh this is like—' and everyone says, 'yes, yes, that's true.' However, I have noted that those who have done not one, but two or three tastings begin to say, 'No, not me, I do not smell this, rather I smell that.' This is a great result, it means that you are already becoming critical about taste, you are growing. The more people are curious the more they go into it, and in turn often our members propose new products to us . . . and so this leads to a constructive dialogue with our experts.

In Cossu's view, sensory knowledge built people's confidence in their own opinions and led them to be more assertive about choosing quality food: taste education led to agency and action.[10] She valued the development of an informed and critical approach to tasting. It was a key strategy of food activism, because it led consumers to become more purposeful about their food, to think about eating and shopping, and to support good local products. Cossu explained: "The senses are the protagonist … It is a continual fibrillation with stimuli coming from several points of view. And that means growing together, synergistically. And hence we work on this. To me, taste is central."

But did it lead to changed behavior? How extensively? For how long? These are questions that are difficult to answer because it is hard to get follow-up data from tasting events and even if people report change we do not know to what extent taste was responsible. I have some evidence from my interviews in Cagliari as well as those I did in 2009 with thirty-eight Slow Food chapter members in several regions of Italy. These food activists unanimously reported that they had become more thoughtful about food and at least some of the time changed their shopping and eating behaviors towards local food from small producers—like the Selargius capers.

For example, in an interview Slow Food Cagliari member Alessandra Guigoni told me that being part of Slow Food was "a positive experience. Before, I didn't think about taste. I began to discover the pleasure of food and of taste from being part of Slow Food." Through Slow Food she became more careful and "more selective" about her food purchases. She used to shop at the supermarket and buy any old products, but Slow Food influenced her to shop locally in her Cagliari suburb of Elmas from the farmers who sold produce from their doorsteps.

Another Slow Food Cagliari organizing committee member, Carla Marcis, spoke of the need for constant re-education to make changed behaviors stick:

> I have periods when I am stronger and purer, more attentive about my purchases and other times—from laziness or who knows?—my shopping is less 'good, clean and fair.' Hence it makes sense to have tastings but to repeat them over time because at first you are enthusiastic and you go and buy bread and greens locally, and other times you are tired and you go to the supermarket and get everything. So the lessons need to be repeated to restimulate commitment … Otherwise the impact of a tasting can last the first week, the first month, but after that it's rare.

Marcis identified an important caveat about the effect of critical taste education—once and done probably will not bring about lasting change. But repeated experiences can build a cumulative effect, emotional commitment to local producers, and a supportive community that can reinforce transformed shopping and eating.

This exploration of Slow Food's caper food community anniversary celebration shows how taste was defined and redefined through exchanges and interactions in social, public settings. The event did not impose one hegemonic taste but rather encouraged the development of critical perception and reflexivity through the involvement of mind, body, senses, and community in learning and attunement. It revealed that taste education provides a powerful context for developing embodiment and agency, and for promoting the unification of "knowing, … understanding, … feeling" that Gramsci (1975: 451) deemed essential to oppositional consciousness and political activism.

Notes

1 Thanks to Jim Taggart, Susanne Højlund and Penny Van Esterik for their helpful comments on earlier drafts of this chapter. Thanks to Professor Gabriella Da Re and Professor Benedetto Meloni for invitations to be a visiting professor at the University of Cagliari in 2011, 2015 and 2016, which made this research possible. Deep thanks to the generous and committed people in Cagliari's alternative food sector who participated in the research.

2 For scholarship on food activism in Italy, see Fonte 2013; Grasseni 2013; Guigoni 2014; Leitch 2003; Miele and Murdoch 2002; Orlando 2011; Pratt and Luetchford 2013; Rakopoulos 2014; Sassatelli and Davolio 2010; and Siniscalchi 2013, 2014a, 2014b.

3 See Black 2017 for a discussion of how all the senses contribute to taste and Sutton (2001: 99) on Greeks' particular focus on "auditory qualities of food" in their expression "listen to that smell."

4 Here is the original Gramsci (1975:451–452): "Passaggio dal *sapere* al *comprendere* al *sentire* e viceversa dal sentire al comprendere al sapere . . . non si fa storia-politica senza passione." I thank Gianni Pizza (2013: 89) for highlighting this citation and for his fine work on Gramsci's relevance to anthropology; see also Counihan 1986.

5 On Slow Food see Andrews 2008; Sassatelli and Davolio 2010; Siniscalchi 2013, 2014a, 2014b; and www.slowfood.com accessed October 25, 2017.

6 See http://slowfood.com/international/8/slow-food-terminology accessed October 14, 2015.

7 Slow Food has celebrated taste and "the right to pleasure" since its founding in 1983 as Arci-Gola—from "ARCI" (*Associazione Ricreativa e Culturale Italiana*), the leftist Italian association for recreation and culture (Siniscalchi 2014a: 75), and "*gola*," meaning desire for and love of food (Counihan 1999: 180).

8 Slow Food teaches about taste and its relationship to local, organic food by holding farm visits, tastings, and "Master of Food" courses on key products like wine, cheese, olive oil, and coffee. On Slow Food's Master of Food courses see www.slowfood.it/educazione/master-of-food-slowfood/ accessed May 13, 2017.

9 On the Slow Food presidium caper from the island of Salina see www.fondazioneslowfood.com/it/presidi-slow-food/cappero-di-salina/ accessed June 23, 2017.

10 Anna Cossu's observations are supported by those of anthropologist Nicolas Sterndorf Cisterna (2014: 93): "tasting is a bodily practice that can for the most part be developed, and indeed often times connoisseurs report their ability to taste to be enhanced as they become more experienced."

References

Andrews, G. (2008). *The Slow Food Story: Politics and Pleasure.* Montreal: McGill-Queens University Press.

Black, R.E. (2012). "Wine Memory." *Sensate Journal.* http://sensatejournal.com/2012/06/rachel-black-wine-memory/ accessed February 7, 2017.

Black, R.E. (2017). "Sensory Ethnography: Methods and Research Design for Food Studies Research," in J. Chrzan and J. Brett, eds., *Food Culture: Anthropology, Linguistics, and Food Studies, Research Methods for the Study of Food and Nutrition* vol. 2, 228–238. New York: Berghahn.

Bosco, F.J. (2006). "The Madres de Plaza de Mayo and Three Decades of Human Rights Activism: Embeddedness, Emotions, and Social Movements." *Annals of the Association of American Geographers*, 96(2): 342–365.

Bosso, C., ed. (2016). *Feeding Cities: Improving Local Food Access, Security, and Resilience*, New York: Routledge.

Bruner, J. (1986). *Actual Minds, Possible Worlds*. Cambridge, MA: Harvard University Press.

Cisterna, N.S. (2014). "Unexpected Moments and the Wine Experience." *Food and Foodways* 22 (1–2): 90–111.

Counihan, C. (1981). *Food Culture and Political Economy: Changing Lifestyles in the Sardinian town of Bosa*, doctoral dissertation, Anthropology, University of Massachusetts.

Counihan, C. (1984). "Bread as World: Food Habits and Social Relations in Modernizing Sardinia." *Anthropological Quarterly*, 57(2): 47–59.

Counihan, C. (1986). "Antonio Gramsci and Social Science." *Dialectical Anthropology*, 11(1): 3–9.

Counihan, C. (1990). "La ricerca sul campo in Sardegna: un'odissea professionale e personale." *Quaderni Bolotonesi*, 16: 43–52.

Counihan, C. (1999). "The Body as Voice of Desire and Connection in Florence Italy," in *The Anthropology of Food and Body: Gender, Meaning, and Power*, 178–194. New York: Routledge.

Counihan, C. (2004). *Around the Tuscan Table: Food, Family and Gender in Twentieth Century Florence*. New York: Routledge.

Counihan, C. (2014a). "Cultural Heritage in Food Activism: Local and Global Tensions," in R. Brulotte and M.A. Di Giovine, eds., *Edible Identities: Exploring Food and Foodways as Cultural Heritage*, 219–230. Farnham: Ashgate.

Counihan, C. (2014b). "Women, Gender, and Agency in Italian Food Activism," in C. Counihan and V. Siniscalchi, eds., *Food Activism: Agency, Democracy, Economy*, 61–76. Oxford: Bloomsbury.

Counihan, C. and V. Siniscalchi, eds. (2014). *Food Activism: Agency, Democracy, Economy*. Oxford: Bloomsbury.

Fonte, M. (2013). "Food Consumption as Social Practice: Solidarity Purchasing Groups in Rome, Italy." *Journal of Rural Studies*, 32: 230–239.

Gramsci, A. (1975). *Quaderni dal carcere*, vol. 1. Torino: Einaudi.

Grasseni, C. (2013). *Beyond Alternative Food Networks: Italy's Solidarity Purchase Groups*. Oxford: Bloomsbury.

Guigoni, A. (2010). "Il cappero di Selargius. Aspetti storici e culturali di una pianta ultracentenaria." *Anthropos e Iatria*, 14(4): 8–13.

Guigoni, A. (2014). "Ciborami contemporanei e mestiere contadino. Tra resistenza e omologazione. Il caso di studio Sardegna," in G. Turus, ed., *Bioresistenze: cittadini per il territorio: agricoltura responsabile*, 87–104. Padova: Esedra.

Howes, D. and C. Classen. (2013). *Ways of Sensing*. Florence, GB: Routledge.

Leitch, A. (2003). "Slow Food and the Politics of Pork Fat: Italian Food and European Identity." *Ethnos*, 68(4): 437–462.

Le Lannou, M. (1941). *Pâtres e Paysans de la Sardaigne*. Tours: Arrault e C.

Marovelli, B. (2014). "Meat Smells Like Corpses: Sensory Perceptions in a Sicilian Urban Marketplace." *Urbanities* 4(2): 21–38.

Meneley, A. (2014). "Discourses of Distinction in Contemporary Palestinian Extra-Virgin Olive Oil Production." *Food and Foodways*, 22 (1–2): 48–64.

Miele, M. and J. Murdoch. (2002). "The Practical Aesthetics of Traditional Products: Slow Food in Tuscany. *Sociologia Ruralis*, 42(4): 312–328.
Morini, G., D. Luiselli, D.S. Risso, S. Tofanelli. (2015). "Il gusto degli Italiani," *La Cultura del Cibo*, vol. III, *Il cibo in Italia*, 365–375. Torino: UTET.
Ochs, E., C. Pontecorvo and A. Fasulo. (1996). "Socializing Taste." *Ethnos* 61 (1–2): 7–46.
Orlando, G. (2011). "Sustainable Food vs. Unsustainable Politics in the City of Palermo: The Case of an Organic Farmers' Market." *City and Society* 23(2): 173–191.
Pink, S. (2009). *Doing Sensory Ethnography*. London: Sage.
Pizza, G. (2013). "Gramsci e de Martino. Appunti per una riflessione." *Quaderni di Teoria Sociale*, 13: 75–121.
Pratt, J. and P. Luetchford. (2013). *Food for Change: The Politics and Values of Social Movements*. London: Pluto Press.
Rakopoulos, T. (2014). Food Activism and Antimafia Cooperatives in Contemporary Sicily," in C. Counihan and V. Siniscalchi, eds., *Food Activism: Agency, Democracy, Economy*, 113–128. Oxford: Bloomsbury.
Sassatelli, R. and F. Davolio. (2010). "Consumption, Pleasure, and Politics: Slow Food and the Politico-aesthetic Problematization of Food." *Journal of Consumer Culture*, 10(2):1–31.
Scarpellini, E. (2004). "Shopping American Style: The Arrival of the Supermarket in Post-War Italy." *Enterprise and Society* 5(4): 625–668.
Seremetakis, N. (1993). "The Memory of the Senses: Historical Perception, Commensal Exchange and Modernity." *Visual Anthropology Review*, 9(2): 2–18.
Siniscalchi, V. (2013). "Environment, Regulation and the Moral Economy of Food in the *Slow Food* Movement." *Journal of Political Ecology*, 20: 295–305.
Siniscalchi, V. (2014a). "La politique dans l'assiette. Restaurants et restaurateurs dans le mouvement Slow Food en Italie." *Ethnologie française*, 44(1): 73–83.
Siniscalchi, V. (2014b). "Slow Food between Politics and Economy," in C. Counihan and V. Siniscalchi, eds., *Food Activism: Agency, Democracy, Economy*, 225–241. Oxford: Bloomsbury.
Sutton, D.E. (2001). *Remembrance of Repasts: An Anthropology of Food and Memory*. New York: Berg Publishing.
Taggart, J.M. (2012). "Interpreting the Nahuat Dialogue on the Envious Dead with Jerome Bruner's Theory of Narrative." *Ethos*, 44(4): 411–430.
Teil, G. and A. Hennion. (2004). "Discovering Quality of Performing Taste: A Sociology of the Amateur," in M. Harvey, A. McMeekin and A. Warde, eds., *Qualities of Food*, 19–37, Manchester: University of Manchester Press.

13

Reindeer Fat and the Taste of Place in Sámi Food Activism

Amanda S. Green

Introduction

It is late December and I have been at my field site in Jokkmokk, Sweden, for over three months. In the deep cold of winter, my research on Sámi food activism and food practices was underway. The sun set two weeks before and we would not expect to see it for another three weeks. Still, life was bustling in the small town. During the holiday rush, I assisted at the reindeer meat butcher shop where my husband worked as they were busy with the holiday season demand.

During a lunch break our supervisor's father invited us over for a special stew. We threw our snowy shoes and heavy jackets in a side room off the front door. We grabbed seats at a table fully set with a pitcher of *saft* (lingon berry juice), potatoes, *knäckerbröd* (hard bread), and butter. On the stove, a pot of reindeer broth, meat, and tongue simmered. Our supervisor pulled out a packet of thin northern bread and told us, "This is the appetizer." He dipped the bread into the pot. The previously dry, hard bread was limp with broth. I did the same. It was delicious and fatty, but as it cooled the fat hardened and began to coat my mouth and my plate. This I found less appealing. I immediately learned that reindeer fat is distinctively solid at room temperature. It's best eaten when very warm.

Several months later I invited another butcher over for dinner in order to learn more about his work. I admitted to him that I found reindeer fat to be weird. I told him, "The first time I ate it, I panicked because it was all over my mouth." He laughed and related to me that his grandmother used to add lots of reindeer fat to the blood sausages she would make. He admitted that he hated it when he was a child, yet all of the adults said they loved it. He explained, "Now I do like it. I don't know what happened. It's like beer. You learn to like it as you grow up."

In between these two personal moments during field research, reindeer fat emerged as an important theme in Sámi food activism. I attended several meetings and conferences where the unique taste and nutritional value of reindeer fat was celebrated. For example, in March I participated in a meeting between members of the national Agriculture Board and several of Jokkmokk's food producers. The project leader for the

Sámi Culinary Center, a culinary education program, presented to the gathered audience. Using a set of slides, she showed us the unique nutritional composition of reindeer fat. She explained, "There isn't much research on reindeer meat. We took some samples from the fall slaughter, the fat, liver, and kidneys. We compared the reindeer fat with butter, as butter is what we often cook with." The results of her research on reindeer fat were featured in numerous presentations and on the pages of cookbooks, websites, and reports. They all shared some variation of the following statement:

> New research gives support for the old Sámi knowledge of the reindeer's good fat. Measurements from 2013 show that reindeer fat from the stomach and back have significantly lower cholesterol levels (61 mg) than the most commonly used animal fats, including butter (256 mg), lard and tallow. Iron values are over 100 times higher in the fat and even Vitamins B9 and B12, riboflavin, thiamin, and zinc are significantly in higher concentration (Slow Food Sápmi n.d.).

Following the presentation, a government representative asked her about the taste of reindeer fat. She responded that it had an umami taste, a savory taste hard to describe. Throughout that year, I heard reindeer fat described as having an umami taste, not having a gamey taste, or having an unusual taste. A Jokkmokk chef I interviewed explained that reindeer have special fat because they live in such a cold climate where they burn their fat and grow it back every year. This chef was the first to tell me that reindeer fat is rock hard when it is in cold water and it has a high burning point, making it ideal for cooking. From observations, I found that people typically procured reindeer fat either from their own or friends' reindeer or from small boutique butchers in order to make sausage, stews, or to cook with. It was not a standard ingredient at the grocery store.

Based on these observations and taste experiences, I was surprised that reindeer fat emerged as a focus in Sámi food activism. Reindeer fat is not necessarily the easiest ingredient to procure or taste to appreciate. As evidenced in my friend's account, children are socialized into enjoying the taste of reindeer fat. Yet taste preferences are also socialized, or cultivated, at other key moments in our lives, established most recently in the works of Paxson (2012) and Weiss (2012) who demonstrate how and why consumers wish to acquire a palate for or discernment of particular tastes. Certain types of consumers acquire social capital by demonstrating their connoisseurship and cultivated palate through their discernment of *terroir* (Paxson 2012: 192). Or, as Brad Weiss notes, certain consumers seek to cultivate a palate for particular (or even peculiar) tastes in order to reconfigure their relationship to place (2012: 453). Similar processes are occurring between Sámi activists and interested Swedish and Sámi consumers. In this chapter, I suggest that Sámi activists are using the French concept of *terroir*, roughly translated as "taste of place," to reach consumers interested in cultivating a taste of place.

The *terroir* concept has emerged as a global term to discuss the importance of place to the production of taste (Trubek 2008: 12). As the discourse of *terroir* became integrated into Sámi and Swedish social worlds, I argue that the taste of reindeer fat emerged as a topic in Sámi food activism because the taste of reindeer fat operates strategically as an index of place. *Terroir* discourses enable Sámi activists to describe the relationship cultivated between reindeer herders, reindeer, and the plants of Scandinavia's Arctic landscapes that produce the taste of reindeer fat. This, in turn, allows activists to call attention to the landscapes that are threatened by encroaching extractive industries and repercussions of climate change, moving the taste of reindeer fat into a discussion of Sámi rights to food sovereignty. In this chapter, I address the following questions: Why is the taste of place produced in reindeer fat coming to matter? How is the taste of reindeer fat produced by place, and how are Sámi food activists building public awareness of this link between place and taste? How is this taste of place expanding the (political) discourses and frameworks available to discuss the nature of Sámi rights?

Methods and Background

I have been conducting ethnographic fieldwork on Sámi food activism since 2008. I participated in or observed over 100 food-centered events and contributed to food workshops, worked in butcher shops, attended food conferences and observed reindeer slaughter courses. My work as a community researcher for the Jokkmokk Food Assessment also brought me into planning meetings and discussions with food producers and activists where reindeer fat was concerned.

The term Sámi describes the indigenous peoples historically and currently living in Sápmi, the territories of the Sámi people that extend across Finland, Norway, Sweden, and the Russian Kola Peninsula. There are approximately 15,000 to 20,000 Sámi in Sweden. Sweden claimed Sámi lands in 1886 and granted a minority of Sámi the right to herd reindeer and to hunt, fish and timber on these lands. Over the next century, Sámi people experienced forced assimilation and further land loss. Yet by 1993, after a century of Sámi activism, all individuals who identify as Sámi have rights to self-determination as Sámi which are recognized in the Sámi Parliament that represents Sámi interests to the national government (Beach 2007). Still, rights to use and maintain lands granted to certain Sámi remain largely unrecognized and unclear (Anaya 2011; Lantto and Mörkenstam 2008; Henrikson 2008). Indigenous sovereignty, including Sámi food sovereignty, is inherently geopolitical: colonialism is at its core about dispossession of land (Simpson 2011; Rifkin 2009). The taste of reindeer fat reflected in *terroir* indexes or points to violations of Sámi sovereignty when their ability to produce this taste of place is limited because a lack of access to land, caused by action (or lack of action) by the Swedish state.

This research took place primarily in Jokkmokk, a 19,477 square kilometer municipality with 5,000 inhabitants (see Figure 13.1). Jokkmokk was chosen as my primary research site because it is a Sámi cultural and political center. Resources like the *Samernas Utbildningscentrum* (Sámi Folk School) create a space for inculcating cultural competence and knowledge for Sámi individuals (Olofsson 2004). The annual Jokkmokk Winter Market as well as Jokkmokk's network of hiking trails, rivers and lakes attract tourists seeking Europe's "last wilderness," which is actually the living, herding, fishing, gathering and hunting grounds for many Sámi from the area. The

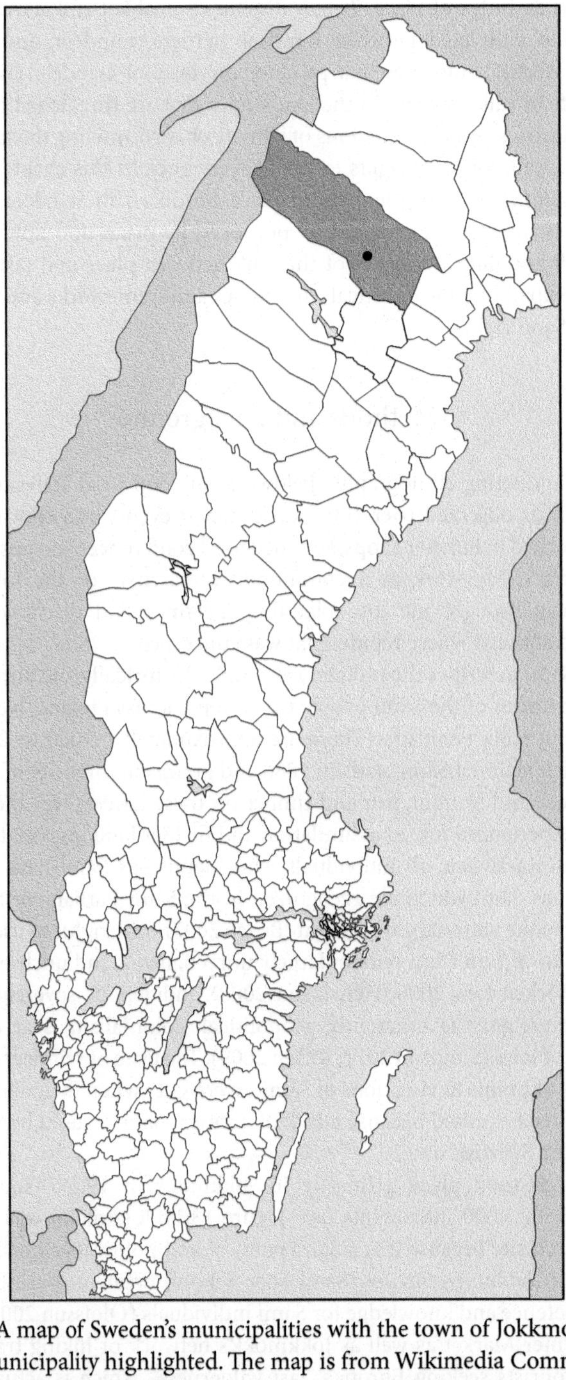

Figure 13.1 A map of Sweden's municipalities with the town of Jokkmokk and Jokkmokk Municipality highlighted. The map is from Wikimedia Commons licensed under Creative Commons and is by Fred J at the userpage: http://commons.wikimedia.org/wiki/User:Fred_J

region hosts incredible cultural and natural diversity at the UNESCO World Heritage Site *Laponia*, chosen in recognition of the biological diversity of the Swedish mountains and the transhumance practiced by Sámi reindeer herders.

Jokkmokk's history is one of settlement and colonization of Sámi lands in order to exploit natural resources, including timber, hydropower and iron (Össbo and Lantto 2011; Brännlund and Axelsson 2011; Baer 1994). Jokkmokk was also chosen as a research site because mining companies have been prospecting for mineral resources and applying for permits to begin development of an iron ore mine just outside of Laponia.[1] Laponia and the national parks have been critical to the preservation of the remaining land needed for local reindeer herding, fishing, hunting and gathering. Food activists from the region frame the mine as a choice the Swedish government must make between an unsustainable mining industry or a sustainable, local food system, as one research participant rhetorically asked, "Is it a mine or fish in Sápmi?"

The contrast in rural development possibilities for Jokkmokk—mines or local food systems—became particularly stark during my research when it was named Sweden's 2014 Culinary Capital, as part of Sweden's national program to build Sweden into the New Culinary Nation (*Nya Matlandet*). Government agencies billed the program as an effort to revitalize Sweden's rural regions through the development of artisanal foodways and the increase in food exports and tourism. During my tenure in the remote municipality of Jokkmokk, key national programs came to have very local impacts as mines and local food production were offered as visions of the future. Sámi cultural and culinary practices were highly valued in the discourses of many programs, but Sámi rights to access and manage the land from which these prized culinary practices arise remained weak and under constant threat by other industries and interests. These contemporary circumstances, as well as the historic conditions of Sámi rights and recognition in Sweden, laid the groundwork for a taste politics of reindeer fat to emerge.

Literature Review

In this chapter, I use the term "taste" to refer to a unique molecular-level and sensory quality of reindeer fat which, in turn, is being identified, commodified and politicized by Sámi food activists in order to advance their cultural, political and economic projects. I understand the taste of reindeer fat to exist within the relationship between "the human and the material" (Højlund 2017, citing Teil and Hennion 2004). The taste of reindeer fat comes into being as Sámi food activists define and make political the material taste qualities of reindeer fat, taste qualities that have gone unobserved and unarticulated until this moment. Indeed, the taste of reindeer fat is both inherently material and inherently social, its qualities arising in the intersection of the fat's molecular composition and this sociopolitical moment where its particular molecular composition (i.e. taste) comes to matter.

Fat is a complex taste quality to define. Researchers have sought to categorize the taste of fat as a sixth taste receptor in addition to salty, sweet, sour, bitter and umami.

This new taste, titled *oleogustus*, (a combination of "oil" and "taste" in Latin) stimulates specific taste bud receptors through fatty acids (Running et al. 2015). Fatty acids (the molecules responsible for the taste of fat) in fact taste rather bad with bitter-ish and sour-ish tones which researchers believe may have evolved as an indicator that food is rotten. However, in taste tests, participants tend to identify the taste of fat with its "mouthfeel," the creaminess and viscosity of fats, which is the result of triglycerides rather than fatty acids. From these sensory studies of fat, researchers have observed that participants are largely unable to articulate the taste of fat. Thus, researchers suggest that there may be a limited vocabulary to describe the sensation of the taste of fat (Singh 2015; Keast and Costanzo 2015).

Throughout my research, the descriptions of reindeer fat were vague, primarily referring to a rich taste and a good taste but giving little other detail. Instead, participants explained fat's nutritional value and gave an order or preferences for types of fat based on its location (back fat, kidney fat, caul fat, etc.) and harvest season (fall or winter, primarily). When attempting to articulate the taste of reindeer fat myself, I was only able to conjure up descriptions of bad taste experiences when the fat was rancid (a rare but memorable occasion), confirming both the bad taste of accumulated fatty acids and the difficulty in articulating the taste of fat (though according to many, older Sámi enjoyed the taste of oxidized fat). In this fat-vocabulary vacuum, the taste of reindeer fat is in fact "in the making" (Højlund 2017) in the sense that Sámi activists are calling our attention to a taste that might be appreciated but is perhaps difficult to articulate.

According to Amy Trubek, "taste evaluations must occur through language, through a shared dialogue with others" (2008: 7). In the case of fat, this language is being developed both by the *oleogustus* researchers as well as food activists. This calling of attention to fat's taste occurs within the growing importance of *terroir* to public articulations of taste. *Terroir*, a historically French term, is currently used globally to describe the taste a product like wine or cheese might accrue as the product of a specific landscape, climate, and cultural tradition. Through *terroir*, according to Trubek, certain foods are understood to embody and express a place including the material conditions of production as well as the cultural knowledge of production.

In the context of Sámi foodways, this taste of place emphasizes the taste of fat as a taste that comes from the reindeer's diet of lichens, herbs, grasses and other wild plants that the reindeer grazes upon as she migrates from Scandinavia's subarctic forests to mountainous pastures. This is apparent in excerpts from the 2014 cookbook released by Slow Food Sápmi entitled *Smak pa Sápmi*:

> Reindeer fat is a culinary gold with high nutritional content and rich tastes. The reindeer builds up her fat reserve over the summer and uses it during the winter. The fat gets its good taste and composition from the feed which is composed of grass, plants, lichens, reindeer lichens, crowberries and other wild plants.
>
> Harnesk 2014: 3

In this example, the sensory taste qualities of reindeer fat are coming to be defined in the framework of a taste of place, a discourse of *terroir*, a term that I found was occasionally invoked by Sámi food activists. Paxson notes that *terroir* is both descriptive and prescriptive (2012: 188). It may be a descriptive label that adds value by promoting a product's place-based distinction as often done in the United States. It may be prescriptive when producers agree to terms of production that are then regulated by systems such as the geographic origin labels of the EU. In Sápmi, the taste of place experienced through reindeer fat is strategically used to describe an indigenous foodway and assert rights to those foodways. This deployment of *terroir* reflects Paxson's observation that "*Terroir's* appeal lies precisely in its ideological flexibility; it can be translated to frame various relations between place and production" (Paxson 2012: 211).

This research adds another layer to the descriptive and prescriptive scaffolding by describing how *terroir*, when reinterpreted and deployed by indigenous peoples in the EU, can become an assertion of indigenous rights. *Terroir* is a descriptive label which, by highlighting the material and cultural conditions necessary to the production of appropriate reindeer fat, can assert indigenous rights to the entire set of practices necessary to the production of fat. The production of tasty reindeer fat is more than a question of taste but a question of how valuing that specific taste can benefit Sámi society by expanding the discourses and frameworks available to discuss the nature of Sámi rights. The taste of reindeer fat comes to matter in the intersection of the *terroir* concept and the recognition of Sámi rights to their lands.

An Ecology of Reindeer Fat Production

The unique adaptation of reindeer and reindeer herders to the Scandinavian Arctic environment are what give reindeer fat its unique taste of place and nutritional profile. These adaptations, their importance to cultural practice and knowledge, and the limitations imposed on these practices by Swedish settler colonialism are also what make reindeer fat a galvanizing symbol for Sámi political activism. The taste and health of reindeer fat reflects a specific reciprocal management relationship developed between reindeer herders, reindeer and regional ecosystems over several centuries. This ever-changing set of relationships can be termed an "ecology of production," a term which calls attention to multispecies activity (reindeer, pastures and herders) that contribute to food production within the particular social, economic and political forces (Paxson 2012: 32) that influence Sámi reindeer herding.

Contemporary reindeer herding takes place across nearly 40 percent of Sweden, with reindeer migrating from the boreal forests of the Baltic coastline to the alpine mountains along the Norwegian border (see Figure 13.2 for an example of alpine pastures in Laponia). While the reindeer herding of the twentieth and twenty-first century is vastly different from the original reindeer herding which evolved in the fifteenth and sixteenth centuries, reindeer continue to graze in much the same way (Forbes et al. 2006; Sköld 2010; Müller-Wille et al. 2006: 377). They typically graze on the vegetation of native pastures which includes grasses and shrubs in the summer mountain pastures and lichens and smaller shrubs in the winter forest pastures. The varied diet of the reindeer leads to a varied consumption of fat and fatty acids which contribute to the higher omega-3 fatty acid composition of reindeer fat (Wiklund et al. 2001).

Figure 13.2 Reindeer graze on the summer alpine pastures in the Laponia UNESCO World Heritage Site, Jokkmokk, Sweden, in August 2014 (photo by Amanda Green).

Reindeer are sometimes given pelleted feeds (a combination of oat, wheat, bran, sugar beet pulp and soy bean meal) in order to add more weight before slaughter or to supplement when natural grazing is limited, particularly during harsher winters. Weight increases from supplemental feedings are a win for the reindeer owner and the meat industry, as owners earn more money from heavier carcasses and reindeer get better carcass scores indicating better quality meat according to industry standards (Wiklund et al. 2000). However, in the study of meat composition, researchers find that feed type affects the quality of an animal's fat, especially the fatty acid composition. This pattern is also responsible for the nutrient composition of reindeer meat (Wiklund et al. 2003). Wiklund and co-researchers found differences in the nutritional composition among reindeer grazed for two months exclusively on natural pastures or a commercial pelleted feed mixture (Wiklund et al. 2001). Natural pasture grazing reindeer had higher fatty acid compositions. This in turn impacts the taste and health of reindeer meat and fat.

Regarding taste, participants in studies have reported no noticeable differences in the sensory properties between lichen-fed and commercial-fed reindeer meats (Wiklund et al. 2000, 2003). Yet, there is considerable consensus among reindeer herders that there is variation in reindeer flavor and tenderness based on pre-slaughter handling, supplementary feeding with commercial feed, and seasonal variation in diet. Indeed, such consensus also reflects broader claims by my research participants that certain herders can identify their own reindeer by taste because of the specific composition of their grazing grounds. In many ways, Sámi food organizers aim to reshape consumers' taste experiences to mimic their own by educating them in the hopes of elevating their sensory palate to this level. Educated consumers may come to discern and prefer the taste of naturally grazed reindeer over commercially-fed reindeer (an effort similar to the goals of North Carolina local pork advocates described by Weiss 2011).

Researchers suggest that, in order to reap the most benefits from consumer interest in healthy fat and meat quality, commercial feeds should more closely mimic natural pastures by including a higher amount of fatty acids found in natural grazing grounds, such as linseed oil (Wiklund et al. 2001: 297). Another, more radical, proposal would be to ensure that reindeer and reindeer herders have consistent access to and management of grazing grounds so that their reindeer can graze on natural pastures year-round. This production of reindeer meat would require a readjustment of the meat industry to value the taste and nutritional profile of reindeer grazed exclusively on natural pasturage over the heavier carcass weight of pellet-fed reindeer. It would also require an adjustment from consumers to come to prefer the taste—as well as the health—of naturally grazed reindeer. I turn now to a discussion of *gurpi* to illustrate how these connections between place and taste are constructed by Sámi food activists.

Cultivating the Connection between Place and Taste: The case of *Gurpi*

Food became a clear target of Sámi political activism in the early 2000s when Slow Food Sweden invited Sámi reindeer meat producers to join the organization. The first Slow Food Sweden Presidia, or sponsored food community, was actually a Sámi product, *suovas*. The *Suovas* Presidia featured a lightly salted and cold smoked reindeer meat. The impact of the Presidia on Sámi participants was clear: the Presidia organized Sámi reindeer processing companies where there previously had been no organization and raised Sámi expectations about their own cuisine (Åhren 2013). By 2009, a separate Slow Food Sápmi Convivia was established to promote Sámi autonomy over their culinary products.

In 2013 Slow Food Sápmi began to establish its second Presidia product: *gurpi* (see Figure 13.3). *Gurpi* is a fresh sausage made from reindeer meat that is coarsely chopped and then wrapped in caul fat and smoked. The importance of caul fat to the production of *gurpi* is a key reason fat has emerged as a discussion point in Sámi food activism. *Gurpi* is consistently described as a food for when one "works in the woods" with the reindeer. A butcher I worked with explained that *gurpi* should be fried and eaten immediately. If it gets cold, the fat hardens and it is less tasty. *Gurpi* is culturally significant: children learn to prepare *gurpi* from their teachers at the public Sámi Elementary School in Jokkmokk. A school teacher explained that the caul fat is a difficult ingredient for them to find when they prepare *gurpi*, but they always get it because it's very good fat.

A party was held by Slow Food Sápmi in Umea in 2014 to celebrate the release of their new cookbook *Taste of Sápmi*.[2] Hosted in the Sámi-inspired Tráhppie Café located near Umea's public museum, the event brought together Slow Food Sápmi

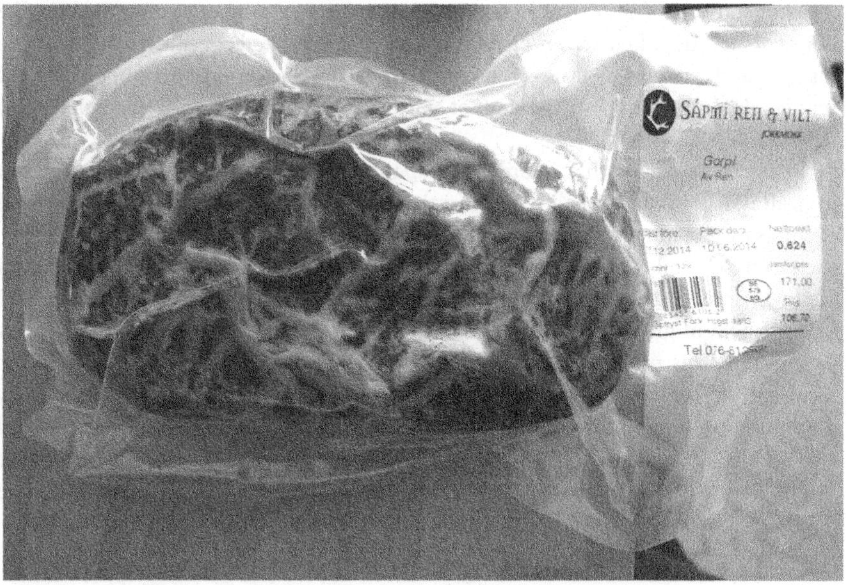

Figure 13.3 *Gurpi*, pictured here, is made from ground reindeer meat that is wrapped in caul fat and smoked. Sápmi Ren och Vilt, a wild game butcher shop in Jokkmokk, sold *gurpi* under the label gorpi (photo by Cate Dolan).

board members and participants, foodies, and the press to celebrate Sámi foods. The walls were adorned with the book's photographs and tables were spread with appetizers prepared from the book's recipes, including *gurpi*. *Gurpi* was a major headliner for the event. I listened to a presentation of *gurpi* by a chef from South Sápmi. He explained to the audience, "When we make *gurpi* we use the stomach fat, the stomach net. Or, it's called the *aisalaken* in South Sámi. (. . .) When you're out in the forest, working with the reindeer, you take bread, usually the *gahkku*, and the *gurpi*." His narrative tied together the reindeer, South Sámi language and history, and the woods of southern Sápmi. Such narratives locate *gurpi* in particular places and people: the South Sámi pastures, reindeer and the reindeer herders who are speakers of the South Sámi language. At this event, we were able to taste *gurpi* with knowledge of its production and purpose in Sámi culture. We could then purchase it from local butchers on site.

During our final weeks in Jokkmokk, my husband and I were invited to the autumn slaughter of the bull reindeer. On a crisp autumn day, we climbed into the hills where the entire herd was gathered to separate the males for slaughter. The blueberry leaves had turned red, and the clear skies allowed us to see into Sarek National Park. As the slaughter began the following morning, several individuals prepared to document the production techniques for *gurpi* for educational and marketing purposes. A professional photographer and reindeer herder filmed the traditional process of removing the caul fat from the reindeer (see Figure 13.4). I was told that the filming was undertaken as part of the documentation and revitalization efforts as well as the marketing of reindeer

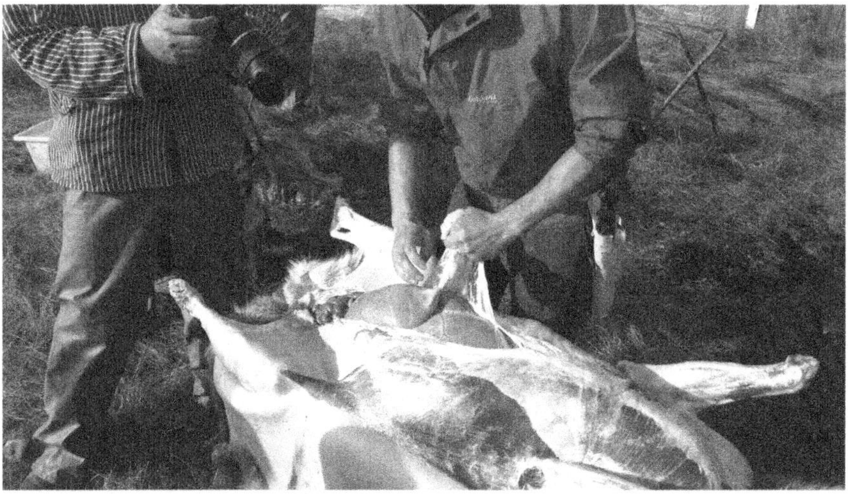

Figure 13.4 During the annual fall slaughter of reindeer, a herder-butcher removes the caul fat from the reindeer while another reindeer owner-photographer documents the process (photo by Amanda Green).

meat. The film, released in April 2017, is primarily a marketing resource for Renlycka, a third-party certification for Sámi-owned reindeer, to teach consumers about this unique food product (Renlycka 2017). During the documentary, one follows the reindeer from the open pastures to the corrals with the reindeer herders, to the slaughter, butchery, and kitchen table. A narrative is recorded in North Sámi and subtitled in English. This production locates reindeer fat in particular places (the beautiful pastures we witnessed on our climb to the corral) and with particular people (the herders in the pastures and the butchers at the corral and butchery). Thus, as in the example above, it accomplishes the connection between the taste of *gurpi's* reindeer fat, place, and cultural practices.

Challenges to the Taste of Place in Reindeer Fat

Not all reindeer can be raised within the particular reciprocal management relationship discussed in this chapter because of particular social, economic and political forces. Within the context of Swedish settler colonialism—the nexus of settlement, land seizures, hydropower, timbering and mining projects in the north of Sweden—the landscape has been so altered that reindeer and herders have limited access to and control over viable land, which in turn limits the natural grazing availability (Svensson 1987). Furthermore, climate change alters weather patterns making it even more difficult for reindeer and herders to replicate that management relationship which produces healthy and tasty fat from Sámi lands (*The Economist* 2016; Forbes et al. 2016). According to most climate models, temperatures will continue to freeze and thaw during the winter which will make northern Sweden's snow pack wetter and heavier. The reindeer will be unable to access the natural winter feeds that ensure the quality of their meat and fat.

Renlycka's president explained the threat to a group of herders during a meeting I observed in June 2014: "If you put a reindeer on pelleted feed for more than two or three months, the meat is changed. The vitamins decrease, the minerals decrease, and the fat becomes dangerous fat. The system depends on natural feed." Putting reindeer on pelleted feed alters the fat's nutrition and it breaks the *terroir* link between the reindeer's taste and the places on which it has grazed. Yet, this is not a simple decision for reindeer owners. Rather it intersects directly with land access, climate change and economic decision-making.

I witnessed this during the 2013–2014 winter when warm temperatures and heavy, wet snow combined to create a thick layer of snow-ice concrete. The reindeer were no longer able to break through the snow pack to get to the critical winter feed that comprise their winter diet. Most herders gathered their reindeer in corrals and fed combinations of pelleted feed, hay, and in some cases, tree lichens. During a slaughter course this same winter, a butcher showed the students the caul fat around the stomach. He explained to us that it was best to remove it now during the slaughter process. He removed the caul fat and spread it across the deep snow. The students collected it later when they prepared to process the different parts of the reindeer. There was also a reindeer whose fat was the wrong color and its stomach was distended. The butcher told us that it had been on pelleted feed too long. He instructed the students not to take its fat.

With a warming Arctic climate and with increasingly limited access to grazing lands, the current marketing of reindeer as a healthy meat and fat with a taste reflecting its place of production may become increasingly challenging for most reindeer owners to produce. Herders will have to use pelleted feed rather than natural pastures. The end result is that the selection of reindeer fat as part of marketing to consumers becomes potentially risky for herders whose reindeer fat no longer tastes of place.

Yet, the focus on reindeer fat's taste of place creates an opening for the integration of land rights into Sámi food activism. Specifically, herders must have access to land in order to produce flavorful and nutritious reindeer fat that reflects the *terroir* of its production. This turn towards land in food activism is not surprising given that it was land which nation-states such as Sweden took from native peoples (Moreton-Robinson 2007). The specter of losing more land haunts Sámi food organizing.

Constructing a discourse of the *terroir* of reindeer fat enables food activists and producers to assert their rights to land and culture in a new discourse. In the careful construction of this discourse, the potential exists for a message that is counter-hegemonic in its criticism of unsustainable uses of land, misrecognition of indigenous land rights, and the differential impacts of climate change. Crafting marketing and consumer education around the taste of place enables Sámi food organizers to position Sámi rights within broader public concerns over the consumption of authentic tastes and authentic places.

Swedish settler colonialism continues to dismantle Sámi food systems and interfere with Sámi food sovereignty: their capacity to adapt their food system to contemporary ecosystems and climates as well as their capacity to reproduce their material and social culture (Whyte 2014; Kirsch 2001). At the intersection of the indigenous food sovereignty framework and the discourses of *terroir*, an agenda emerges for recognizing indigenous rights to produce their foods. The growing recognition and importance of *terroir* extends this framework to recognize the importance of the set of reciprocal management relationships between land, water, animals, plants, and humans (Raster and Hill 2016) that are necessary to the production of nutritious reindeer fat. Recognizing the importance of place to the production of reindeer fat's taste and nutritional value leads to a recognition of the larger ecology of production—the relationship between managed grazing pastures, reindeer, and herders that produce healthy and tasty fats as well as the political and economic structures which shape the ways herders can operate. Threats to those relationships are threats to food sovereignty and to the original rights established through treaty, in the case of the Sámi, through the Reindeer Grazing Acts and the 1751 Lapp Codicil.

Conclusion

Reindeer fat and the taste of place come to matter because of an assemblage: globally circulating concepts of *terroir* and food sovereignty meshed with a mobilized indigenous collectivity focused on food activism. Reindeer and the landscapes they occupy are iconic of Sáminess, and *terroir* offers a means for activists to deploy this symbolism. Through documentaries and public discussions of reindeer fat that link together the taste experience of products made from reindeer fat like *gurpi* with Sámi languages, landscapes, and cultural practices, organizers begin to create the knowledge and discourse necessary for the construction of a *terroir* discussion of reindeer fat. The language to describe the taste of reindeer fat is still in its nascent stage, and time will tell if a well-articulated taste of fat vocabulary will emerge, perhaps through the work of the *oleogustus* research advocates.

Given the linguistic vacuum that exists to describe the taste of fat, I suggest that nutrition science provides an alternative discourse through which we can know reindeer fat. Remembering that fatty acids are the molecular structure which scientists find may be the sixth taste, tracking fat's fatty acid composition is a way to discuss the alteration of reindeer fat by different feeding practices. Indeed, nutrition science makes the invisible visible (Kimura 2016: 124) by making the nutrient content of reindeer fat grazed on natural pastures visible. This in turn makes visible the importance of reindeer fat's *terroir*, its taste of place, which comes from the natural pastures. Nutrition science data thereby provide evidence of the alteration of a culturally significant animal and food source. These (now visible) changes to reindeer fat indicate that Sámi food sovereignty may have been violated by thwarting Sámi abilities to control their own food system.

The discourses of *terroir* as well as nutrition science enable a different conversation to take place about Sámi rights by making visible a different reality, one which acknowledges the importance of land to the production of appropriate reindeer fat. Without these discourses and the science and vocabulary they make available, we would be having a different or having no conversation about indigenous food sovereignty involving taste and fat. Instead, we can talk about an ecology of production, or a set of reciprocal management relationships, that recognizes the landscapes, plants and animals and cultural practices that produce a taste of place and more nutritious reindeer product.

Notes

1 In October 2014, the Swedish agencies charged with evaluating mining applications determined that a mine in the region would significantly impact the ongoing livelihoods of the area (particularly the herding of reindeer) and no mining projects will be permitted at this time (Sveriges Radio 2014).

2 The presentation from the book release are available to stream at this website: http://urskola.se/Produkter/181224-UR-Samtiden-Tema-Samisk-mat.

References

Åhren, C. (2013). "Sámiska sociala entreprenörer: En kamp för ett folks självbestämmande, självförtroende och överlevnad" [Sámi social entrepreneurs: A fight for a people's self-determination, self-confidence and survival]. SESPA: Mittuniversitet.

Anaya, J. (2011). "The Situation of the Sámi People in the Sápmi Region of Norway, Sweden and Finland." UN General Assembly.

Beach, H. (2007). "Self-Determining the Self: Aspects of Saami Identity Management in Sweden." Acta Borealia 24: 1–25.

Forbes, B.C., M. Bölter, L. Müller-Wille, J. Hukkinen, F. Müller, N. Gunslay, Y. Konstantinov, eds. (2006). *Reindeer Management in Northernmost Europe*. Springer.

Forbes, B., T. Kumpula, N. Meschtyb, R. Laptander, M. Macias-Fauria, P. Zetterberg, M. Verdonen, A. Skarin, K. Kim, L. Boisvert, J. Stroeve, A. Bartsch (2016). "Sea Ice, Rain-on-Snow and Tundra Reindeer Nomadism in Arctic Russia." *Biology Letters* 12(11): 20160466.

Henriksen, J. (2008). "The Continuous Process of Recognition and Implementation of the Sámi People's Right to Self-Determination." *Cambridge Review of International Affairs* 21(1): 27–40.

Højlund, S. (2017). "Taste," in *Wiley: The International Encyclopedia of Anthropology*, in press.

Keast, R. and A. Costanzo (2015). "Is Fat the Sixth Taste Primary? Evidence and Implications." *Flavour* 4(5): 1–7.

Kimura, A. (2016). *Radiation Brain Moms and Citizen Scientists.* Durham: Duke University Press.

Kirsch, S. (2001). "Lost Worlds: Environmental Disaster, 'Culture Loss,' and the Law." *Current Anthropology* 42(2): 167–198.

Lantto, P. and U. Mörkenstam (2008). "Sámi Rights and Sámi Challenges: The Modernization Process and the Swedish Sámi Movement, 1886–2006." *Scandinavian Journal of History* 33(1): 26–51.

Moreton-Robinson, A., ed. (2007). *Sovereign Subjects: Indigenous Sovereignty Matters*, Sydney, Australia: Allen and Unwin.

Müller-Wille, L., J. Hukkinen, F., Müller, M., Bölter, B.C. Forbes, (2006). "Synthesis: Environmental and Sociopolitical Conditions for Modern Reindeer Management in Europe's North," in B.C. Forbes, M. Bölter, L. Müller-Wille, J. Hukkinen, F. Müller, N. Gunslay, Y. Konstantinov, eds, *Reindeer Management in Northernmost Europe*, 365–379. Springer.

Olofsson, E. (2004). "In Search of a Fulfilling Identity in a Modern World: Narratives of Indigenous Identities in Sweden and Canada." Dissertations in Cultural Anthropology vol. 2. Uppsala: Uppsala University.

Össbo, Å. and P. Lantto (2011). "Colonial Tutelage and Industrial Colonialism: Reindeer Husbandry and Early Twentieth-century Hydroelectric Development in Sweden." *Scandinavian Journal of History* 36(3): 324–348.

Paxson, H. (2012). *The Life of Cheese: Crafting Food and Value in America*. Berkeley: University of California Press.

Raster, A. and C. Hill (2016). "The Dispute over Wild Rice: An Investigation of Treaty Agreements and Ojibwe Food Sovereignty." *Agriculture and Human Values* 1–15.

Renlycka. (2017). "*Gurpi*." www.youtube.com/watch?v=2cO5yPSXSUg, accessed May 16, 2017.

Rifkin, M. (2009). "Indigenizing Agamben: Rethinking Sovereignty in Light of the 'Peculiar' Status of Native Peoples." *Cultural Critique* 73: 88–124.

Running, C., B. Craig, and R. Mattes (2015). "Oleogustus: The Unique Taste of Fat." *Chemical Senses* 40: 507–516.

Simpson, A. (2011). "Settlement's Secret." *Cultural Anthropology* 26(2): 205–217.

Singh, M. (2015). "Salty, Sweet, Sour. Is it Time to Make Fat the Sixth Taste?" *The Salt*. www.npr.org/sections/thesalt/2015/02/18/386100165/salty-sweet-sour-is-it-time-to-make-fat-the-sixth-taste, accessed May 16, 2017.

Sköld, P. (2010). "Development, Adjustment and Conflict: The Sámi and Reindeer Husbandry in Sweden in the Light of Political, Social and Economic Changes," in H. Antonson and U. Jansson, eds., *Agriculture and Forestry in Sweden Since 1900: Geographical and Historical Studies*, 475–491. Stockholm: The Royal Swedish Academy of Agriculture and Forestry.

Slow Food Sápmi. (n.d.) "Ingredients in Sámi Cooking." http://en.slowfoodsapmi.com/ingredients.html, accessed May 16, 2017.

Svensson, T. (1987). "Industrial Developments and the Sámi: Ethnopolitical Response to Ecological Crisis in the North." *Anthropologica* 29: 131–148.

Teil, G. and Hennion, A. (2004). "Discovering Quality or Performing Taste?" in M. Harvey et al., ed., *Qualities of Food*, 19–37. Manchester: Manchester Univertsity Press.

The Economist. (2016). "On Slimmer, on Shorter! Reindeer." *Espresso*. https://espresso.economist.com/11704817e347269b7254e744b5e22dac, accessed May 16, 2017.

Trubek, A. (2008). *The Taste of Place: A Cultural Journey into Terroir*. Berkeley: University of California Press.

Weiss, B. (2011). "Making Pigs Local: Discerning the Sensory Character of Place." *Cultural Anthropology* 26(3): 438–461.

Weiss, B. (2012). "Configuring the Authentic Value of Real Food: Farm-to-Fork, Snout-to-Tail, and Local Food Movements." *American Ethnologist* 39(3): 614–626.

Whyte, K. (2015). "Indigenous Food Systems, Environmental Justice, and Settler-Industrial States," in M. Rawlinson and C. Ward, eds., *Global Food, Global Justice: Essays on Eating under Globalization*, 143–166. Newcastle-upon-Tyne: Cambridge Scholars Publishing.

Wiklund, E., L. Johansson and G. Malmfors (2003). "Sensory Meat Quality, Ultimate pH values, Blood Parameters and Carcass Characteristics in Reindeer (Rangifer tarandus

tarandus L.) Grazed on Natural Pastures or Fed a Commercial Feed Mixture." *Food Quality and Preference* 14(7): 573–581.

Wiklund, E., A. Nilsson, and B. Åhman (2000). "Sensory Meat Quality, Ultimate pH values, Blood Metabolites and Carcass Parameters in Reindeer (Rangifer tarandus tarandus L.) Fed Various Diets." *Rangifer* 20(1): 9–16.

Wiklund, E., J. Pickova, S. Sampels and K. Lunström (2001). "Fatty Acid Composition of M. longissimus lumborum, Ultimate Muscle pH Values and Carcass Parameters in Reindeer (Rangifer tarandus tarandus L) Grazed on Natural Pasture or Fed a Commercial Feed Mixture." *Meat Science* 58: 293–298.

14

Political Taste

Inclusion and Exclusion in the Slow Food Movement

Valeria Siniscalchi

After their final meeting of 2010, the national board members of Slow Food Italia were invited to dinner in the town of Alba, in the Piedmont region of Italy, near the movement's international headquarters in Bra. From the beginning of my fieldwork a year earlier, I had attended all of the board's meetings as an observer, or "the spy" as I was referred to by some of the members and leaders of Slow Food, and this position landed me an invitation among the guests at that dinner. All the guests were requested to bring a bottle of wine, and of course considering they were connoisseurs even if no one worked directly in the wine sector, it would have to be an excellent bottle.[1] I was somewhat bewildered by the task but finally settled on a "natural" French wine with an exotic note that I thought might compete with the other wines. Moreover, the producer's name brings to mind that of a very expensive Piedmont wine, and gave me the opportunity to play with a certain ambiguity when I announced my contribution. Over the course of the dinner, at least eighteen bottles were opened, their nose savored, and the range of flavors and characteristics analyzed in a multitude of exchanges and commentaries.

In *The Taste Culture Reader*, Caroline Korsmeyer points out that "Tastes are subjective but measurable, relative to culture and to individual, yet shared; fleeting sensations that nonetheless endure over many years in memory; transient experiences freighted with the weight of history. And finally, tastes can provide entertainment and intellectual absorption" (2005: 8). But taste can be political too. In this chapter, I delve into this unique space created by taste and practices of tasting in order to analyze the evolution of the political use of taste by Slow Food over the years. From the creation of the association in the mid 1980s, taste has been one of elements at the heart of Slow Food.

The association is a readily visible, well-known entity that has provoked both strong criticism and unconditional adhesion, producing abundant literature about itself and becoming highly mediatized in Italy, its country of birth, and in other countries such as Germany, the United States, and Switzerland where it has a strong presence. This heavy coverage in the media produces an optical illusion of false familiarity when in fact, its internal workings are not at all transparent to outsiders, and not even to its own members. The Slow Food "machine" is extremely complex, combining different aspects and components. It is both an international association composed of various national structures with nearly 100,000 members worldwide, and a widely supported movement which has in turn inspired other "slow" movements. Composed of private enterprise as well as a non-profit structure, Slow Food has created a private university, the *Università di Scienze gastronomiche*. Like other forms of mobilizations in the field of "food activism" (Counihan and Siniscalchi 2014), Slow Food evolved over the course of its existence, expanding its actions and becoming a legitimate actor in the political arenas of food production and consumption as well as other food-related issues such as climate change and the menace to common goods like resources, energy and biodiversity. I have approached and examined Slow Food as an economic and political object that sometimes acts as an institution but one whose actions are conducted through a variety of actors, primarily activists and volunteers.

This chapter is based on my extensive fieldwork inside Slow Food's international and national headquarters in Bra and in some of its political hubs, where for three years I explored the intimate functioning of the movement:[2] the places of power and the ways power is managed, the production of adhesion and trust, the relationship between professionalization and commitment, the imagining of new economic models, and the sometimes slippery compromises with the Market.[3] But before gaining access to the headquarters and the "black box" of Slow Food, I started my fieldwork by tasting, drinking, and eating with members in France and Italy. I also interviewed people who had been excluded or progressively marginalized by the association. This work revealed specific practices of inclusion and less visible practices of exclusion that characterize this movement, its internal dynamics and its ways of functioning. Through the aspects of taste and tasting, I analyze these dynamics and the tension that exists between inclusion and exclusion inside Slow Food.

Taste as a New Way of Commitment

The first Slow Food association was founded in 1986 in the Northern Italian winegrowing region of Langhe. It began as a branch of the ARCI and was first given the name of *Arcigola*. ARCI is the acronym of the *Associazione Ricreativa e Culturale Italiana*, the national organization of leftist and anti-fascist clubs created in 1957. Between the end of the 1970s and the beginning of the 1980s, several thematic associations were created inside the ARCI. They rapidly became autonomous structures known as *Legambiente*, *Arcidonna*, *Arcigay* and *Arcigola*. The *arci* in the name *Arcigola*, is both a reference to the mother organization and an Italian prefix which reinforces adjectives in a superlative manner. The name is completed with *gola*, which literally means gullet, but can also refer to gluttony. At the beginning this was an association of gourmets and consumers devoted to good food and local culinary traditions. The founders and leaders came from the Italian *circoli*[4] that included left-wing political parties, militants and intellectuals who, during the seventies, had been involved in local politics, the trade union movement or local, cultural associations. They were looking for new means and spaces of engagement, in this period of *riflusso* (disenchantment) that characterized the 1980s in Italy, when the ideals and causes that had led large numbers to commit and mobilize collectively seemed to have vanished and ceded to individualism.

The founders of Slow Food reintroduced notions such as "pleasure," that had been rejected during years of social tensions and political battles in the 1970s. This pleasure of drinking and eating well, of conviviality, became one of the pillars of Slow Food's philosophy. But importantly, they speak of a pleasure that is not individual but is a collective and shared pleasure. This is one of the more original ideas of Slow Food: the notion of pleasure acts as the unifying element between the previous period of collective political engagement and the ensuing years of disenchantment when the engagement undertaken by the early adherents assumed new forms.

Indeed, Slow Food's founding manifesto dedicated the movement to the right to pleasure, *Manifesto dello Slow Food, Movimento per il diritto al piacere* (Slow Food Manifesto. Movement for the right to pleasure) (Portinari 1987), was published in the monthly cultural supplement, *Gambero rosso*, of the communist independent newspaper, *Manifesto*. At the time, the supplement was edited by the *Arcigola*, and the manifesto was written in ironic prose (although some of that irony may be lost in translation): it declared that this epoch was born in an industrial civilization dedicated to speed, where man has elevated the machine as the ideal model for life, reducing mankind to a species on a path to extinction in a monstrous sort of autophagy. The manifesto warns of the virus of *fast* and all its collateral effects, and suggests that an assurance of suitable doses of sensual pleasure and slow, lasting enjoyment may preserve us from the contagion of the multitude mistaking frenzy for efficiency. It contends that taste and the pleasure of the *gourmandise* must be returned to the table.

This irony in the manifesto is characteristic of the publications and actions of Slow Food which express it in a very articulate vocabulary. The philosophy itself is firmly situated in the Italian culture of the time, particularly in the epoch's intellectual and critical left-wing politics, and it can be seen in diverse forms, both cultural and political, and with the advent of Slow Food, gastronomic. It also inversely reminds us of the manifesto of futurism written by Filippo Tommaso Marinetti and published in *Le Figaro* in 1909. Among other things, Marinetti glorified "aggressive movement" and the "beauty of speed." By contrast the "slow" is opposed to the "fast" in the Slow Food Manifesto which underlines the importance of taking time as a reaction to the frenzy of the "fast life" and the standardization of food and taste. Over time, this first manifesto has been rewritten, adapted, simplified, and translated. But minor variations aside, the various versions all emphasize the wealth of taste and "material" pleasures of the table: "A firm defense of quiet material pleasure is the only way to oppose the universal folly of Fast Life . . . Our defense should begin at the table with Slow Food. Let us rediscover the flavors and savors of regional cooking and banish the degrading effects of Fast Food" (Slow Food Manifesto, cf. www.slowfoodusa.org/manifesto). Extending this perspective, a few years later, *Il dizionario di Slow Food* (the Slow Food Dictionary) defines taste as "the recognition of flavors, expanded to the rituals of the table, stemming from the artistic and intellectual heritage of a civilization (translation)" (Ruffa and Monchiero 2002: 87–88). Taste is tied to knowledge and the table, and in a larger sense to the civilization of which they are a part.

Tasting as a Collective and Inclusive Practice

The pleasures of the table and the rediscovery of the taste of regional cooking also became the motor of Slow Food's diffusion into the various Italian regions: Carlo Petrini, one of the leaders and founders of Slow Food and today president of Slow Food international, and his collaborators crisscross the Italian peninsula in search of local products and producers, and *osterie* (Siniscalchi 2014a). The sense of this word has evolved over time. Previously, an *osteria* was thought of simply as a convivial place serving wine and a quick bite to eat, but Slow Food has contributed to an update and revalorization of the term, notably through the guide published by the association each year: *Osterie d'Italia, Sussidiario del mangiarbere all'italiana* (Osterias of Italy, The guide to eating and drinking Italian). Osterias are defined by *Il dizionario di Slow Food* as "places synonymous with traditional cooking, simple service and hospitality without pretention, and quality wines, all for a reasonable price (translation)" (Ruffa and Monchiero 2002: 118).

These restaurateurs often become ambassadors of good will for the association and their osterias, proposing regional cooking with local products, become gathering places for members and followers of the movement: "I operated a restaurant, the *Ochina Bianca*, for thirteen years. One day each month, I closed the restaurant and it became a place for Slow Food events, and I organized dinners and meetings of the association" (Gilberto V. 12/9/2009). As both a restaurateur and leader of the local Slow Food group in the town of Mantova, Gilberto is one of the "elders" of the association. His case certainly illustrates the blend of gastronomy and collectivism, but it also points to certain informal policies inherent in the association. The get-togethers that he organizes at his establishment—dinners or tastings—are a means of supporting the existence and visibility of Slow Food, and marking territory through a grass-roots presence of the association. They also serve to enlarge the circle of members and followers in an atmosphere of conviviality imbued in the sharing of wine and dishes of local cuisine.

These events, organized around a product or a wine and also open to non-members, transform tasting into a collective practice and at the same time, a means of inclusion. At the end of each event or get-together, unaffiliated participants are encouraged to purchase a membership card. These methods of expanding the association and spreading its principles have not been limited to Italy and the early years of the association. As Slow Food has spread across diverse portions of the globe, the methods have been tailored to blend with local customs in each country, and they have come to characterize the history of Slow Food's evolution (Siniscalchi 2017).

Although taste and tastings facilitate the recruitment of new members in the association, they also define and demarcate the membership in Slow Food: these events are opportunities to meet friends, to exchange and share practices and knowledge about food in the atmosphere of an extended family, in essence, characteristic of the osterias where they often take place. But above all, the tastings are mechanisms for collectively shaping taste, and developing membership through the appreciation of taste. They contribute to the creation of a type of "cultural intimacy" (Herzfeld 1997) of taste that becomes a distinctive characteristic of the association's membership.

Playing with Taste, Training Through Taste

The capacity of these practices based on taste to define or mark adherence to the group is particularly evident in one particular event organized by the Italian structure: the *Gioco del piacere* (Game of pleasure). In June 2010 the employees of Slow Food Italia in Bra, primarily those from the association office in charge of relations with the members and local groups (*condotte*, in Italian, *convivia* in other countries), were preparing to work late into the evening for a tasting game. There were ten people in the office. Cases of olive oil had previously been shipped to all the *condotte* that had decided to participate in the game. The cases consisted of a collection of four different olive oils produced in the Italian regions most recommended by the *Guida agli extravergini* (Guide to extra-virgin olive oils), one of the guides published by Slow Food that presents and ranks a list of high-quality extra-virgin olive oils produced in various Italian regions. The labels had been covered with a ticket bearing a number. The same four bottles, with their original labels, were placed on the table in the association office: one bottle of oil from Puglia, one from Sicily, one from Lazio and one from Liguria. Two of the employees returned with loaves of bread, and the game began—simultaneously—in all the *condotte* registered to participate. In Benevento in the Region of Campania, in Scandicci near Florence, in Verona, and in all the participating local groups, the members gathered in the osterias for a blind tasting of olive oil. In each *condotta* the leader (*fiduciario*)[5] or a knowledgeable member of the group guided the tasting. The point of the game is to have the members of each participating *condotta* classify the oils and then assemble the local results into a national classification. As each *condotta* finished its tasting, the *fiduciario* telephoned the central office in Bra to report the results. In the Bra office, Alberto, Serena, Fabiana, Marco and Fabrizio then averaged the ratings from the various groups and established the best olive oil of the evening.

While waiting for the results from the *condotte* we began to taste the olive oils. Color, then nose, then palate. Each of us expressed our thoughts and preferences, then tasted again and decided on a rating. From time to time the telephones rang with demands for more information from latecomers, and with time to spare, we tasted again and the conversation turned to stories about various *condotte* or comments about certain *fiduciari* that apparently embody the stereotypes of one or another city or group. As the atmosphere relaxed there were a few jokes between employees. Someone decided to call one of his co-workers secretly from an adjoining room, passing for a *fiduciario* and asking pointed questions about the oils and their organoleptic characteristics that no-one could answer. Everybody laughed. Finally the *fiduciari* began to call and everyone returned to their telephones to note the arriving results. The early results were from *condotte* in the North of Italy, some were surprisingly on time or in advance while others were late to finish. With each call and rating, the tension of the game filled the office.

Approaching 11 o'clock in the evening, the last group finally reported and the tasting game ended. Once the ratings were tabulated, the *condotte* were called with the results, identifying the producer of each of the four olive oils and the final national classifications. The *fiduciari* were then able to pass the information and classification to the local members. Each convivium tasting was followed by a dinner and at the end of the evening each participant received a copy of the *Guida agli extravergini*. As noted by certain organizers, the game presents the olive oil tasting and discovery in a convivial and participatory fashion. People participating in the game are "exposed" to high-quality olive oils and learn not only to appreciate them but also to prefer them.[6] But for the members in the *condotte*, as in the small group of employees in Bra, the game also creates social links and a sense of belonging, not only to the local group, but to the larger family of Slow Food. For an evening, with the intermediary help of a central service, these members are connected and participate at the same time in a unique, collective tasting.

While the game of pleasure brings together the members, across geographical spaces and at the same time, for a tasting transformed into a game, the "Master of Food" project—another Slow Food project in the field of taste education— as its name suggests, was conceived as actual training sessions to provide instruction in Slow Food principles by improving knowledge about food and drinks. The website of Slow Food Italia (www.slowfood.it) also emphasizes this need for education. The site affirms that education in taste is the best defense against poor food quality and the best means to combat the standardization of our meals. Initiated and managed by the education office of the association at its headquarters in Bra, and conducted by Slow Food "trainers," the Master of Food program expresses "*il gusto di saperne di più*" ("the pleasure of knowing more" but also "the taste of knowing more"). The Italian word, *gusto*, suggests the idea of the pleasure of both knowledge and taste that is addressed in this project. During one or several meetings that may take place on two or three different levels, the Master of Food programs offer for-pay training to Slow Food members. They are organized on a thematic basis and thus act as a mechanism to pass on the principles of the association to new members or refine the knowledge of experienced members in a particular domain. The more classic programs are organized around products such as cheese or wine, but new programs have been added over time on themes, such as food shopping, garden products, and cooking without waste. These new themes are in line with the association's evolving interests that have become attuned to environmental issues. However, universal concerns over protection of the environment or reducing waste do not invalidate the knowledge and know-how Slow Food has developed in the domain of the senses and seeks to promote through the Master of Food, "Palestra sensoriale" (sensorial gymnastics). As noted on the Slow Food Italia website, "what better than a sensory gym to train our senses and use them to better evaluate the quality of food. The sight, the touch, the smell, the taste, and even our hearing, are indispensable tools to guide our food choices, ancient instruments that have receded from our consciousness (translation)" (www.slowfood.it/educazione/categorie-master/palestra-sensoriale).

Taste as a Means of Defining Borders

For all of these key moments in the functioning life of the association, tasting is a means of creating conviviality and sharing practices (taste practices above all), while promoting ideologies and developing members' affiliation to the Slow Food "family". But becoming or being a member of Slow Food does not mean entering the "inner circle." The internal organization of Slow Food is extremely hierarchical: a simple membership is distinct from being a volunteer, and among volunteers, those with political roles at the local level (regional or national as well as at the level of each *condotta* or *convivium*, in Italy as well as in the other countries) are part of a more restricted circle which is closer to the upper level of the hierarchy situated in the headquarters in Bra. For these elected leaders, who are often the people organizing tasting workshops, dinners, and master of food programs, the mastery and capability of teaching about taste and its subtleness contribute to an even stronger sense of being on the inside of the movement and a select member of the Slow Food hierarchy.

The elected members of the Italian National Board (*segreteria nazionale*), with whom I started this chapter, are the closest to the core of power in the association. The national board is the organ that includes the presidents of the largest regions in terms of the number of members and elected individuals among the headquarters' leaders. A significant part of the decisions that concern the functioning of the Italian association and its policies are made or validated within this board. On the evening of the closing dinner of their term in office, the exchanges and comments concerning each of the wines presented, and the level of tasting competence exhibited by the members of the board at the table confirmed their place in proximity to the upper echelons of the Slow Food hierarchy. Clearly, one can hardly comprehend a leader in the movement who does not know and appreciate wine. This informal and intimate tasting session highlights and delimits the boundaries of the inner group of Slow Food, and particularly reflects the specific political and cultural intimacy of the group. Although these individuals were "elected," they had first gone through a selection or approval process that involves the highest levels of the Slow Food hierarchy.

This evening revealed other forms of distinction that reflect the permanent tension between inclusion and exclusion which, in my opinion, characterizes life within the movement. The board members present were almost exclusively men, some of them, long-term leaders in the movement. The two women, who as regional presidents had the right to be members of the national board, had relinquished their seat to another male member of their regional board, from the beginning of the board's term. Thus the only woman present, besides myself, was the head of the education office, an elected member from the national headquarters. At this extremely masculine dinner-meeting she brought a white sparkling-wine: this choice of a wine more often associated with the preferences of women accentuated the gender composition of the group and, inside it, a gender distinction through wine.

Still another telling episode took place that evening. At the end of the dinner, once two of the less integrated board members had left, one of the early leaders from Bra made a tour of the table and asked the guests to each contribute twenty euros without any explanation. He then brought out a bottle of "Barolo Monfortino" and began to expound upon its characteristics in a sort of initiation to this great and expensive wine, which we had all just participated in purchasing. It was a bottle from 2001 and the winegrower only vinifies once every four or five years. We began tasting the wine in typical fashion by observing the color in the glass, then appreciating the nose; admiring the aromas continued for a long time and everyone was ecstatic. Finally we evaluated its impression on the palate with small sips, all the while listening to an account of its virtues. The tasting continued as the group shared their appreciation of the wine. At the end of this special event, as we all began to leave, the member of the group responsible for this last tasting approached me, "Spy, do you realize what just happened? This evening you tasted a bottle of Monfortino!" In truth, I had not only discovered and tasted a particular wine, but I had also participated in an intimate moment among this restricted social group at the heart of the movement.

The objectives and ambitions of Slow Food have evolved over time, with a growing need to publicly demonstrate the metamorphosis of a food and wine association into a social and political movement (members are encouraged to become more aware of the production process and the need to support small producers), but the role played by wine has remained important. Teil and Hennion suggest that "taste is a performing activity . . . is an action, not a fact; it is an experience, not an object" (2004: 35). During the evening of that final dinner, the sharing and tasting of a renowned bottle of wine was not only an activity of taste; it was also an example of the means and activities used to delimit this inner group of leaders who are historically important in the association.

Tasting as a Tool for Political Struggles

Certainly tasting events are a means of promoting the association and building membership, but they also enable the showcasing of new themes embedded in the heart of its philosophy, as can be seen in the evolution of the master of food programs. At the end of the 1990s the leaders of the movement began to focus on the environment, now one of the most important elements of Slow Food philosophy (Siniscalchi 2013a). At the turn of the century, and with the beginning of the Terra Madre project in 2004, Slow Food's vision and initiatives shifted. Terra Madre is conceived as a network of food actors and small producers from all over the world. At the same time it is one of the main Slow Food events; once every two years Slow Food organizes an international Terra Madre gathering in Turin, inviting thousands of producers, farmers and fishermen but also researchers to discuss food problems and politics, sustainable production and the future of local economies. The movement increasingly includes producers and is concerned with their living and working conditions, creating projects oriented around "food communities," small producers and local economies. The triad "Good, Clean and Fair"—the movement's slogan since the publication of founder and leader Carlo Petrini's book, *Buono, pulito e giusto* (2005)—summarizes the evolution of Slow Food's philosophy. Good refers to taste and organoleptic qualities of food products; clean refers to the conditions of production and to respect for the environment through reduction or elimination of chemicals products. Finally, the notion, fair, refers to social justice and fair working and living conditions for producers. Along with these changes, education has become a central and strategic element for the movement; education in taste acts as a strategic tool for safeguarding products and species but also as a means of focusing attention on producers' perspectives as part of the process of acquiring knowledge about a product.

Taste laboratories (*laboratori del gusto*) have been offered since the 1990s during Slow Food events organized at the local, national and international level (Salone del gusto, Cheese, Slow Fish, etc.). They are aimed at both members and the general public. Through a sensory experience and work on taste, they enable the discovery of products and relationships between products (e.g. wines or artisanal beers and cheeses), but perhaps even more importantly these tasting labs present an introduction to the philosophy of the movement. And parallel with the movement's evolution, they have become more attentive to the aspects of production. Frequently the laboratories welcome producers who come to present their product. The producers may provide a history and explain production methods while the participants taste and savor the flavors and characteristics of his product. The learning that takes place during these events thus goes beyond the direct organoleptic experience of the product, and encompasses elements such as the conditions of production, choices made by the producer and environmental contexts. The tasting practices become rituals in which food plays both a role of connection, and a means of communication: these rituals produce new forms of sociability, in these cases linking producers and consumers through specific knowledge. Beyond these specific and standardized situations, the direct knowledge of local producers is also a central aspect of the osterias. As described by Roberto from the *Osteria La Campanara*, "When I bring certain products to the table, I am comfortable, I can talk about who made them, their history, and how the producer manages to deliver such a product" (May 16, 2013).

Taste also provides part of the reason for including products in the Slow Food list of labeled products (*presidia*). A "*presidium* of biodiversity" is at the same time a product and a project of valorization and promotion with the objective of protecting "high quality" food production that is considered at risk of disappearing. These products or productions are often endangered because they make little economic sense. The producers may have difficulty making a living from their activity or there may simply be few existing producers. In the *presidia* project, taste becomes a political and economic instrument.

For example, the *Fiore sardo* cheese—a *pecorino* (goat cheese) produced by shepherds in the Sardinia region and now a Slow Food presidium—is considered a high-quality product, but its taste is often too strong for a palate accustomed to industrial standard pecorino cheeses. The production and smoking techniques used to preserve and refine the cheese contribute to its particular taste, which in turn excludes it from some markets and thus increases economic difficulties for producers (Siniscalchi 2013b, 2014b). For Slow Food leaders, promoting the *Fiore Sardo* cheese is both an act promoting taste over the more bland results of industrial food, and one defending small producers against the market logics that marginalize and restrict the viability of their activity. Through taste laboratories and training programs offered to members, Slow Food defends a product, supporting its specific taste and promoting a new kind of "goodness."

But taste is also a tool of exclusion, even in this new political dimension. During the processes of selecting products that are being labeled as *presidia*, staff in the Slow Food Foundation for biodiversity negotiate with producers: the production techniques, the quantities, the place of production, the ingredients, and also taste. Slow Food staff and experts acknowledge that they must learn to appreciate tastes that are different from those with which they are familiar, respecting the principle of a subjective palate and a subjective notion of *buono* (goodness), a principle of relativity that they introduced more recently. At the same time they may ask for changes in some aspects of the production process. In these cases taste acts as criteria for including products in the Presidium project, but it can also be a reason to exclude products and producers. The taste of a product might be too far from Slow Food's idea of goodness, or it might signal that the producers have strayed too far from Slow Food principles (and consequently, products may lose the label that had been previously granted).

In February 2011, Slow Food organized a three-day training program for people in charge of *presidia* in each *convivia*, region or country. After an initial presentation concerning different aspects of *presidia* projects, the training involved some exercises and a comparative blind tasting session. The President of the Slow Food Foundation for biodiversity introduced this session with an explanation: "By tasting, I make a judgment on the quality. Now we must turn that around. The story of the quality needs to be told . . . but the story cannot be told through the experience of a single producer because it does not provide enough assurance of sustainability." With his reference to a story he formally introduced a new element: the narrative, the history of the product. "Blind tasting is no longer sufficient; we have to listen to narratives in order to understand . . . the product's characteristic features need to be explained" (P.S. February 19, 2011). When the tasting session started, he asked the participants to taste three cheeses, then to listen to his narration, and taste again. The exercise was repeated with three apples, and three sausages. In each series of products, there was an industrial product, a good product and a *presidium*. Participants were required to identify the best one from the Slow Food perspective. The first series of cheeses was composed of a raw milk *maccagno* (a Slow Food *presidium*), followed by a *maccagno* from a dairy using added ferments that the foundation President explained, "render the taste and smell homogenous; recognizing them during a tasting requires experience, but you will begin to perceive them." The third cheese was a tomme made with pasteurized milk, and the President noted, "Here you can smell the sharp acidity of the fermentation enzymes and the lack of balance." He continued by explaining that to qualify as a presidium production the animals must be fed with field hay in winter as well as summer. He also outlined the specificities of races, and then summarized, "So, in one cheese, there is the race and the raw milk, in the other there is the race but no raw milk, and in the last there is nothing and we have no idea where the milk comes from." After this final explanation, he pointed to each of the three cheeses, one at a time, and asked the participants to raise their hand if they liked it. These collective exercises functioned as a sort of collective control: participants could make mistakes but they could not say that the industrial product is the *presidium*. Taste, tasting practices, and this kind of collective training in taste are also a way to "test" people, a tool of inclusion and exclusion at the same time.

Tasting as a Tool of a Moral Economy

In the Slow Food world, taste also has a moral characterization: "good" food is also "morally" good because it is produced with respect for the environment and animals.

This characterization applies to *Fiore Sardo* cheese, as well as other *presidia*, but it is also used in the case of the "forgotten" fish found at the core of the event, Slow Fish. Consuming these "morally" good products signifies not only respect for the environment but also for the work of producers, and it implicitly helps them to make a living from their work. Fair working conditions must be guaranteed to all producers, but equally, good food must be accessible to all.

In previous texts (Siniscalchi 2014b) I analyzed an article that appeared in the August 18, 2010 edition of *La Repubblica*, one of the most important Italian newspapers, in which Carlo Petrini, the President of the Slow Food International, denounced industrial Sardinian cheese and defined it as a "bad (*cattivo*) cheese because it behaves badly" (C.P. August 18, 2010). He called for a revolt against the infringement of moral rules, symbolized by this bad cheese which behaves badly with respect to its own region and the people who care about this land, and "with the Sardinian shepherds who created the reputation of this cheese" (ibid.). In recent years, the movement's philosophy and actions have evolved towards a redefinition of the limits of human action and the morality of the economy. "Good, Clean and Fair" become the parameters of a moral economy that is partly an alternative to unbridled free market capitalism, and partly a reform of it. The defense of local economies, of farmers and producers all around the world is aimed at creating "communities" of goals—in a performative way—woven around productive activities and imagined economies (Anderson 1983), directly connected to an ideal past that ignored the damage from anomalies in capitalism, industrial production, and the irresponsible exploitation of resources. The more utopian and inclusive dimension of Slow Food's message[7] aims to bring the poles of the food chain closer, connecting all the actors of local economies in different parts of the world to try and change the food system, in the real world, in part because it clashes with the intimate ways in which Slow Food functions.

Since the beginning, taste and tastings have been at the heart of Slow Food philosophy and political practices. During the 1980s taste was a new form of political commitment for the first generation of leaders but also a way of involving new members inside the movement. During the years, tasting games or trainings have defined and demarcated the membership, hiding internal hierarchies and allowing people to feel that they are members of the big Slow Food "family." At the same time, taste is a tool of distinction and exclusion for people who compose the "inner group": sharing some specific taste practices allows them to build a sort of "taste intimacy" revealing some aspects of the internal way of functioning of the movement.

Gusto—meaning "pleasure" and "taste"—is not only about food, but also the knowledge of food, which today is deemed more and more necessary to the political commitment of Slow Food members. Becoming a moral political tool, taste once again creates boundaries: it allows the exclusion of industrial food or morally bad food, but also excludes producers that do not do "good" work, and people that are not able to recognize "good" food. This moral shifting that appears to be increasingly inclusive by allowing small producers from all over the world to feel that they are part of the Slow Food family, is at the same time visibly exposing the tension that exists between the inclusion and exclusion constantly expressed in Slow Food's practices and values, and in the inner workings of the movement.

Notes

1 The members of Slow Food are considered (and they consider themselves) wine connoisseurs. From the beginning, wine has had an important place inside Slow Food. This fact is linked to the origins of the association: before Slow Food, its founders created an association of lovers of Piedmont's wines (*Libera e Benemerita Associazione degli Amici del Barolo*). Also, the *Guida vini* (Wine Guide), is one of the earliest Slow Food publications.

2 My fieldwork on Slow Food began at the end of 2006. At first I followed the evolution of Slow Food's French national association and the activities of some of its local chapters (called *convivia*). Then, from 2009 to 2014, the main part of my fieldwork took place in the association's headquarters and decision-making center in Bra, which also houses the seats of the Slow Food Foundation for Biodiversity, the Slow Food international association, and the Italian association. Approximately 150–170 employees work in the Bra centers. Thanks to friends of Slow Food who accepted me in the corridors of the headquarters of the movement and put up with my presence with great patience and affection. The time spent with them was always a rich experience, both from the human and intellectual point of view.

3 I have addressed some of these aspects in other texts (Siniscalchi 2013a, b, 2014a, b, 2017, among others). I am currently working on a monograph that will be published by Bloomsbury.

4 In Italy, the word *circoli* refers either to groups of individuals with shared interests and ideas or to the places where these groups gather or meet. *Circoli* are usually organized around themes such as sports, politics, culture or trades.

5 The *fiduciario* (convivium leader) is an individual who has the confidence (*fiducia*) of the president and the national headquarters. He is responsible for marking the physical and social territory of the Slow Food *condotta* through the activities of the association.

6 Other scholars analyze similar processes of transformation of taste and preferences; see Mintz (1996); Terrio (2000); Paxon (2013).

7 This dimension of Slow Food philosophy is close to the environmental perspective of Pope Francis. I cannot explore here the analysis of the links between Slow Food and the Catholic Church that I have explored in other works. Here I simply underline the fact that these links contribute to the reinforcement of the moral dimension assumed by the current message of Slow Food.

References

Anderson, B. (1983). *Imagined Communities*. London: Versus.

Counihan, C. and V. Siniscalchi, eds. (2014). *Food Activism: Agency, Democracy and Economy*. London: Bloomsbury.

Fassin, D. (2009). "Les économies morales revisitées," Annales. *Histoire, Sciences Sociales*, 6: 1237–1266.

Herzfeld, M. (1997). *Cultural Intimacy: Social Poetics in the Nation-state*. New York and London: Routledge.

Korsmeyer, C., ed. (2005). *The Taste Culture Reader. Experiencing Food and Drink*. Oxford, New York: Berg.

Mintz, S.W. (1996). *Tasting Food, Tasting Freedom. Excursions into Eating, Culture and the Past*. Boston: Beacon Press.

Paxson, H. (2013). *The Life of Cheese: Crafting Food and Value in America*. Los Angeles, Berkeley: University of California Press.

Petrini, C. (2005). *Buono, pulito e giusto*. Turin: Einaudi.

Portinari, F. (1987). "Manifesto dello Slow Food, Movimento per il diritto al piacere." *Gambero Rosso, Il Manifesto*, Rome.

Ruffa, G. and A. Monchiero. (2002). *Il dizionario di Slow Food*. Slow Food Editore: Bra.

Siniscalchi, V. (2013a). "Environment, Regulation and the Moral Economy of Food in the Slow Food Movement." *Journal of Political Ecology*, 20: 295–305.

Siniscalchi, V. (2013b). "Pastori, attivisti e mercato. Pratiche economiche e logiche politiche nei Presidi Slow Food." *Voci. Annuale di Scienze Umane*, X: 173–182.

Siniscalchi, V. (2014a). "La politique dans l'assiette: restaurants et restaurateurs dans le mouvement Slow Food en Italie." *Ethnologie Française*, XLIV (1): 73–83.

Siniscalchi, V. (2014b). "Slow Food Activism between Politics and Economy," in C. Counihan and V. Siniscalchi, eds., *Food Activism. Agency, Democracy and Economy*, 225–241. London, Bloomsbury.

Siniscalchi, V. (2017). "Slow Food: les politiques locales d'un mouvement international," in T. Grillot and S. Gacon, eds., *Manger autrement*, 63–85. Paris: PUF.

Teil, G. and A. Hennion, eds (2004). "Discovering Quality or Performing Taste?," in M. Harvey et al., eds., *Qualities of Food*, 19–37. Manchester: Manchester University Press.

Terrio, S. (2000). *Crafting the Culture and History of French Chocolate*. Berkeley: University of California Press.

15

Tasting Displacement: Reflections on Freshness

Joan Gross

Introduction

In *The Physiology of Taste*, Jean-Anthelme Brillat-Savarin wrote, "Taste invites us by pleasure to repair the continual losses brought about by life." Brillat-Savarin was referring to the continual expenditure of energy that has to be replaced by caloric intake, but this quote brings to mind other losses that people experience in life and how taste reminds them of those losses. With over 65 million people displaced from their homes (UNHCR 2016), one common loss is the taste of foods that one grew up with; tastes that are attached to memories of food practices that cannot be continued in new foreign surroundings.[1] I follow Højlund's call to examine taste as a social sense, moving "attention from the privacy of the mouth and the subjective, internal reflections to the public space of sharing the experience of eating" (2015: 2).

In this chapter I explore the taste of freshness in the context of displacement. My curiosity about this topic arose from various ethnographic studies that included Latin Americans, mostly in the United States, but I did not design an ethnographic research project focused on taste. Instead, the issue of taste and freshness bubbled up time and again in each of three ethnographic studies that I conducted (Rural Food Insecurity in Oregon, 2004; Food Activism 2005–2013; Positive Deviance as a Catalyst for Sustainable Food Production and Nutrition in the Andes 2013–2015). Latin Americans' commentary on the lack of freshness in US food also came up in mini ethnographies conducted by my students over the past decade. The fact that this issue arose, unsolicited, in multiple ethnographic encounters in both Spanish and English led me to think about it more carefully. How are fresh or processed foods embedded into national cuisines affecting the taste preferences of people from those places? How does a single taste descriptor, "fresh" refer to a multiplicity of tastes and to the cultural activity of tasting? How does political economy project taste into the public sphere? I will address each of these questions in order.

Acquiring a Taste for National Cuisines

The first time I heard Latin Americans contrast the food they were used to with US food along the dimension of fresh versus processed was when I was researching rural food insecurity in Oregon in 2004. I was sitting in the living room of a Mexican family and when the conversation came around to the differences they encountered with American food, the mother of the family, María, said, "*No está fresca*" (It's not fresh). She said that they chose to live close to the school, so that her children could come home for lunch and not eat the school lunch.[2] I knew that this family had joined the local gleaners group, an organization of low-income people who come together to gather food for themselves and for other low-income people who are not able to procure their own food. I also knew that the other members of the group were upset with this family because they discovered that a lot of the food that they delivered to them in weekly boxes ended up in the trash. (It's a very small town without a lot of privacy.) I was beginning to wonder whether this was a case of culturally inappropriate food causing friction in the newly diversified group. I should explain that in rural Oregon, most gleaners are of northern European ancestry. While the original idea was that they would gather the produce left over in farmers' fields, more often they gather and distribute canned and frozen foods and lots of day old pastries. María was happy to help the group repack the food for delivery, but language barriers made communication too difficult and the group leader stopped asking her to help. Instead, the gleaners delivered boxes of food to her house every week. I asked about her involvement with the gleaners and María replied:

> María: At the gleaners there's lots of food that they bring but we don't like it. Like the milk that they sell that is dry. I have prepared it but the children don't like it. They don't drink it. Because I have always bought them from the gallon, fresh. Then, when I prepare it for them they don't drink it. They feel the taste is different.
>
> Joan: What are some other things that the children don't like?
>
> María: the veggies that are frozen.

Canned soups, frozen vegetables, powdered milk and sweet rolls deemed to be too sweet all found their way into the trash.

María's family's rejection of the new tastes of processed US food was echoed in an interview that a student of mine did with another Mexican woman. She said that her sister-in-law took her to the emergency food pantry when she first arrived, but she was sorely disappointed. She said in English, "it is too much cans, and I don't use cans. Like beans in cans and I say, no. I can buy a big bag of beans and I can eat fresh everything." Certainly, an argument could be made that both dried beans and canned beans are processed, not fresh, but in her mind (and in the opinion of most of her compatriots) there is something much more unnatural about canned products. People who grow beans also dry them and cooking practices are built around the dried product. Canned vegetables belong to a different cooking tradition impacted by the industrial food system.

Latin American migrants to the United States are often shocked by how much industrially processed food is consumed there. In a survey done in 2008 of 106 Latinas over age eighteen in Linn and Benton counties in Oregon, a very strong preference was found for fresh produce over non-perishable food (Patton-Lopez and Dodge-Vera

2008). That same year, in focus groups Oregon State extension faculty conducted with Latino youth in a local high school, many of the teens commented that American food was prepared from cans and boxes and was "not fresh."

These different preferences were noted by volunteers at emergency food pantries who reported that Latin Americans chose fresh vegetables, often avoided by Anglo Americans. One young Anglo woman when offered a winter squash simply said "What is it and what do you do with it?" Of course, the issue is more complicated. Some pantry users have no cooking facilities, refrigeration, or time to cook and therefore stock up on food that they see as being easier to prepare. However, low income Latin Americans do not generally have more facilities and more time than low income Anglos. The difference is better explained by divergent national culinary practices that developed under different political economies to which I will return below.

"Fresh" as a Complex Signifier

"Fresh" is a polysemous term in both English and Spanish and there is considerable overlap in both languages. Both languages use the word to refer to water and air in much the same way and it is also used to modify other nouns, but I am only concerned here with those meanings in which the referent is food. "Fresh" is used to describe the taste of food, but it does not refer to one or even a combination of the five discernable tastes. It incorporates multisensory qualities such as mouth feel, color and smell. Furthermore, the qualities indexed by the term vary according to the product. Fresh arugula is bitter and crisp; fresh strawberries are sweet and soft. We might call "fresh" a gustatory shifter since it parallels linguistic shifters like pronouns. For example, "I" means me when I speak it, when you say it, it refers to you. In a Piercean framework, these are called referential indexes or indexical symbols (Silverstein 1976). The idea of *terroir* in the realm of taste operates in a similar manner. Rather than referring to a particular taste in the mouth, it indexes the soils, climate, and culture of a particular region. It creates a taste of a specific place (Trubeck 2008). If place is the meaningful core of *terroir*, then time is the meaningful core of "fresh." Fresh food by definition is perishable.

However, place also plays a role in freshness because foods do not grow everywhere and people living in different ecosystems have access to different types of fresh food. So, if the element of time is short as the term "fresh" implies, then the connection to the place where the food was grown is implicit. The amount of time necessary to ship products across space cancels the freshness factor and if you grew up picking and eating subtropical and tropical fruits and vegetables, biting into an old mango is tasting displacement.

This brings us to two different meaning clusters of the term "fresh." One way to analyze meaning is to see what the word contrasts with. When Latin Americans malign North American food as not being fresh, the examples that they use to illustrate their point fall into one of two categories. When "no está fresca" is followed by talk of canned vegetables and powdered milk, "fresh" means "unprocessed." In this case, the word contrasts with food in which the original shape, color and taste is transformed and chemicals are usually added for preservation. When "no está fresca" is followed by a description of mangos and avocados bought in US grocery stores that pass from too green to rotten, never becoming delicious, the meaning of "fresh" is "newly harvested." This can be applied to meat as well. One Mexican farmworker interviewed by my student Melva Treviño spoke of the chickens he bought in US grocery stores as having a strange flavor and texture which he attributed to the time between when it was killed and when it was sold. "Back home if you wanted to eat chicken for lunch, you got up early in the morning and grabbed a live one," he told her.

We can refer to these meanings as Fresh_1 (unprocessed) and Fresh_2 (newly harvested). Fresh_1 stands as an ungradable opposite of processed food, whereas Fresh_2 can be located along a gradable scale. In other words, we can say x is fresher than y, but not as fresh as z. This freshness scale is something consumers look for at farmers' markets and grocery stores when they choose their food. In doing so, they use a variety of senses, but they are all merely ways of discovering what is most important—taste. Often shoppers are not allowed to taste before they buy and, besides, many things they would want to cook first anyway, so they must guess at the taste by using visual cues of color and form, tactile cues such as soft spots, olfactory cues indicating newly picked fruit, or rotten meat, and even auditory cues such as the tapping of melons. Employing sight (long considered the master sense in the West) in order to determine taste (considered a subordinate sense) reverses, in some way, the dominant sensory hierarchy in the West (Sutton 2001:4). Of course, there are other concerns that lie outside sensory hierarchies, most obviously, price. Often, the food that feels, smells, sounds, and (especially) looks most appealing is also more expensive. However, even within the bins of discounted foods, people still make selections, and the projection of what the food will taste like is the point of the selection process.

Fresh_2, then, refers to the age of the food, or the time between when it was alive and harvested and when it was eaten. The world is predominantly urban now and under the global industrial food regime, the age of food and where it was grown is often a mystery. The only sure way to know if something is fresh is if we pluck it ourselves from the land that grew it or purchase it from a trusted producer. Here, I draw on my own experiences as a hobby food grower and food forager, but also on research that I conducted in Ecuador on nutritional practices of rural food producers (Gross et al. 2017) to try to understand the shared human experience of tasting food gathered directly from its source. One of my favorite things to do in the summer is to stroll around my garden picking and eating random fruits and vegetables, the taste of blueberries and raspberries fresh off the bush, cherry tomatoes wrapped in a basil leaf, fresh Italian parsley. As I describe this eating event, incorporated in the remembered taste is the warmth of the sun, the smell of the roses and the ripening grapes, the crispness of fresh greens, the colors of berries and flowers, the patterns in Speckled Roman heritage tomatoes.

As I fondly reminisce eating from my own garden, I also remember the pleasure that rural Ecuadorians expressed in having access to fresh foods and herbs when I was conducting ethnographic research there. I see the smile on Margolita's face as she bent over and picked *zunfo,* an aromatic wild plant growing on the side of the path on our walk up to water the bulls. She crushed it in her hands and offered it to me to smell its intoxicating perfume. I hear the children talking to each other about how delicious the milk foam from a freshly milked cow tastes. I remember the pleasure Don Oswaldo took in digging up *mellocos* (*Ullucus tuberosus*) and telling me that I should prepare the tubers for dinner that night and taste how much better they are than ones bought in the store. He once told me, "Potatoes you buy in Quito are five days old or more. It's just not the same." His son had already migrated to Quito and when Don Oswaldo visited him there, he missed the food from his home.

I would like to suggest that in addition to being influenced by the situations and conditions of the activity of tasting as Højlund points out (2015: 3), taste is also influenced by the procurement of the food being tasted. Don Oswaldo was convinced that I would immediately prefer the taste of fresh tubers. What I preferred was the relationality that I had to him and his garden through the food that I ate. My palate was not experienced enough in the physical taste of Andean tubers to perceive a difference. I enjoyed eating fresh *mellocos* because I remembered how Don Oswaldo showed me how to gently lift the earth and pluck the small rosy tubers from the vine. Understandably, if the procurement process consists of undervalued, backbreaking, monotonous work in the hot sun for insufficient wages, that, too, leaves its trace on the taste of the food.

One of the major challenges for the food industry is how to keep food tasting fresh (whatever that taste may be). For rural food producers like Don Oswaldo, the very idea that you can "keep" food fresh is an oxymoron. Talking about the fish being sold from a car in his village, he said "Imagine how many days old they are. I think they refrigerate them so that they don't spoil." This would strike most North Americans as an odd statement because we are trained that keeping food "fresh" is exactly the point of refrigeration. For Don Oswaldo, it was as if they were trying to play a trick on consumers. Refrigeration was masking the age of the fish. Freshness, by his definition, cannot be maintained through time. You can't "lock in freshness" no matter what Tupperware says.

"Fresh," then, may refer to a variety of different tastes and may be opposed to either processed food or old food. Perhaps its already vague, yet always positive meaning makes it such a popular term for marketing foods in the global industrial food system. "Fresh!" or "Fresca!" can be found labeling a variety of foods that do not qualify as either Fresh_1 or Fresh_2. Labeling canned salsa as "fresh" contradicts the central meaning of Fresh_2 because it eliminates the important element of time sensitivity. Labeling ultra-processed instant soups as "fresh" contradicts the meaning of Fresh_1 which contrasts fresh with processed foods. These two examples clearly mark the term as a floating signifier with no agreed-upon meaning.

How does Political Economy Project Taste into the Public Sphere?

Michael Carolan wrote about his preference as a child for canned over fresh mushrooms as an example of how "consumer "preference" is produced and maintained through practice—through literally *doing* those tastes over and over again" (2011: 6). Tastes are not good or bad in their essence. Preferences are acquired over time, influenced by practice and relations, but also by public discourses about food. We can apply the aspect that procurement plays in taste to mushrooms. When I taste mushrooms that I have gathered in the woods, the experience of finding a treasure trove of golden chanterelles popping up through the moss around a decaying trunk in the cool forest forms part of my pleasure in tasting the mushrooms fried in butter with garlic back in my kitchen.

Cuisines and eating practices develop in different ecological zones, with influence from different cultures, but also from business interests concretized in food policies. USDA crop and milk subsidies support the overproduction of cheap processed foods that fill emergency food pantries, school kitchens, and the bellies of Americans, edging out perishable, and labor intensive fresh-foods which become more expensive. It is not surprising then that North American cultural tastes are more oriented towards processed foods that are cheap and easily prepared (Fitzgerald and Petrick 2008). This is particularly true of low-income Americans because food-spending is flexible, while rent, transportation and utilities are mostly fixed. The emergency food system that many low-income Americans depend on is based on surplus processed foods. The taste for these among poor North Americans illustrates Bourdieu's explanation that people learn to want what is possible for them, developing a taste of necessity (1990). A taste for expensive fresh foods and the time required to prepare them can even be seen as elitist by low-income Americans who depend on inexpensive commodity foods distributed by the USDA to feed their families.

Though most Latin American immigrants fall into the low-income category in the United States, they do not share the cultural tastes of low-income North Americans. The transition from fresh to processed foods and from subsistence to industrial farms happened much earlier in the United States, than in Latin America. In Latin America, fresh, whole foods grown locally are usually quite affordable and their processed counterparts in grocery stores are generally more expensive. National cuisines and the taste of necessity developed within them vary depending on ecosystems and what policies are in force. With the introduction of free trade agreements, Mexican corn had to compete with subsidized US corn and many smallholder farmers had to leave their land and the fresh food it produced, migrating to cities and abroad where fresh tastes became a cherished memory (Bacon 2013). The climate is changing, but not as quickly as people are changing climates, moving from places and tastes that they know to places where things don't taste right.

National cuisines also change through time. Since the implementation of the North American Trade Agreement (NAFTA) in 1994, the market share in Mexico for processed foods has been growing by leaps and bounds (along with obesity). McMillan (2015) reported that high fructose corn syrup exports to Mexico are 863 times what they were before NAFTA and sales of American processed food products have tripled since 1999. Walmart which opened its first Mexican store in 1991, operated 2,387 stores in the country as of November, 2016 according to Wikipedia. The immigrants we spoke with grew up in rural areas in Mexico before the 1990s where the nutrition transition to processed foods had not made much headway. It might be that for young Mexican immigrants arriving today, freshness would not be cited as the key difference between Mexican and US food. Not only are Mexicans eating more processed foods, but US residents are eating more fresh food and much of this unprocessed food comes from Mexico (McMillan 2015). Spearheaded by local food activists, the amount of fresh food in Oregon food pantries has gone up considerably in the last decade. What this means is that the unsolicited characterization of US food as "not fresh" in contrast with the food that Latin American residents were used to is not be essentialized, but is rooted in a particular historical moment shaped by different national agrifood policies.

Conclusion

In this chapter, I made a foray into the lived experience of political economy through the taste descriptor "fresh." The ubiquity of both "not fresh" food and Latin American workers in the United States is a consequence of political economic policies, policies that support the agroindustrial food system and that have devastated smallholder farmers on both sides of the border. Laplantine points out that "there exists a political and a historical dimension to sensory experience (2015: 83). Latin Americans in the United States have to confront both processed food and food that was harvested long ago and "kept fresh" through various means. "No está fresca" summarizes the taste of displacement. As an index, "fresh" bears a spatio-temporal contiguity to that which it describes. That is the link between the taste of freshness and one's home where the pleasurable act of eating fresh food repeatedly took place. Food growers around the world have formed their sensorial sphere around the taste of really fresh food. Neoliberal free trade agreements have forced countless numbers of smallholder farmers off their land to become work seekers in national metropolises as well as foreign countries where they have to subsist on strange food. So, to turn the opening quote on its head, it's not the pleasure of taste that repairs the losses in these cases, it's the memory of the pleasure of tastes no longer available that reminds one of the loss of home.

Notes

1 The recognition of this fact and the ways in which cooking and eating together break down barriers between cultures has led to multiple local initiatives such as the Refugee Food Festival sponsored by Food Sweet Food in France, or our local cooking classes at the Corvallis Multicultural Literacy Center.

2 A local community food assessment found that the most exposure that Latin Americans had with North American food was the school lunch program of their children (Patton-Lopez and Dodge-Vera 2008).

References

Bacon, D. (2013). *The Right to Stay Home: How US Policy Drives Mexican Migration.* Boston: Beacon Press.

Bourdieu, P. (1990). *The Logic of Practice.* Stanford: Stanford University Press.

Brillat-Savarin, J-A. (1825/1949). *The Physiology of Taste*, Trans. M.F.K/ Fisher. New York: Heritage Press.
Carolan, M. (2011). *Embodied Food Politics*. Burlington: Ashgate.
Fitzgerald, G. and G. Petrick. (2008). "In Good Taste: Rethinking American History with our Palates." *The Journal of American History*, September, pp. 392–404.
Gross, J., C. Guerrón-Montero, M. Hammer, and P. Berti. (2017). "Creating Healthy Bodies in Rural Ecuador at a Time of Dietary Shift." In *Food, Agriculture and Social Change: The Vitality of Latin America*. Stephen Sherwood, Alberto Arce and Myriam Paredes, eds. New York: Routledge.
Højlund, S. (2015). Taste as a social sense: rethinking taste as a cultural activity. BioMed Central. See https://flavourjournal.biomedcentral.com/articles/10.1186/2044-7248-4-6 https://flavourjournal.biomedcentral.com/articles/10.1186/2044-7248-4-6 (Accessed January 9, 2017).
Laplatine, F. (2015). *The Life of the Senses. Introduction to a Modal Anthropology*. London, New Delhi, New York, Sydney: Bloomsbury.
McMillan, T. (2015). How NAFTA Changed American (and Mexican) Food Forever. February 13, 2015. The Salt, National Public Radio. www.npr.org/sections/thesalt/2015/02/13/385754265/how-nafta-changed-american-and-mexican-food-forever (Accessed November 13, 2015).
Patton-Lopez M. and T. Dodge-Vera. (2008). *Las Comidas Latinas: Community Needs Assessment for Nutrition Education Programming*. Corvallis: Oregon State Extension Service.
Silverstein, M. (1976). "Shifters, Linguistic Categories and Cultural Description," in eds., K Basso and H. Selby, *Meaning in Anthropology*. Albuquerque: University of New Mexico Press.
Sutton, D. (2001). *Remembrance of Repasts: An Anthropology of Food and Memory*. New York: Berg.
Trubek, A. (2008). *The Taste of Place: A Cultural Journey Into Terroir*. Berkeley and Los Angeles: University of California Press.
UNHCR (2016). Global forced displacement hits record high. See www.unhcr.org/en-us/news/latest/2016/6/5763b65a4/global-forced-displacement-hits-record-high (Accessed December 30, 2016).

Index